Du Tarzan *Ich Jane*

Gender Codes im Design | Birgit Weller und Katharina Krämer

You Tarzan *Me Jane*

Gender Codes in Design | Birgit Weller and Katharina Krämer

blumhardt verlag hannover

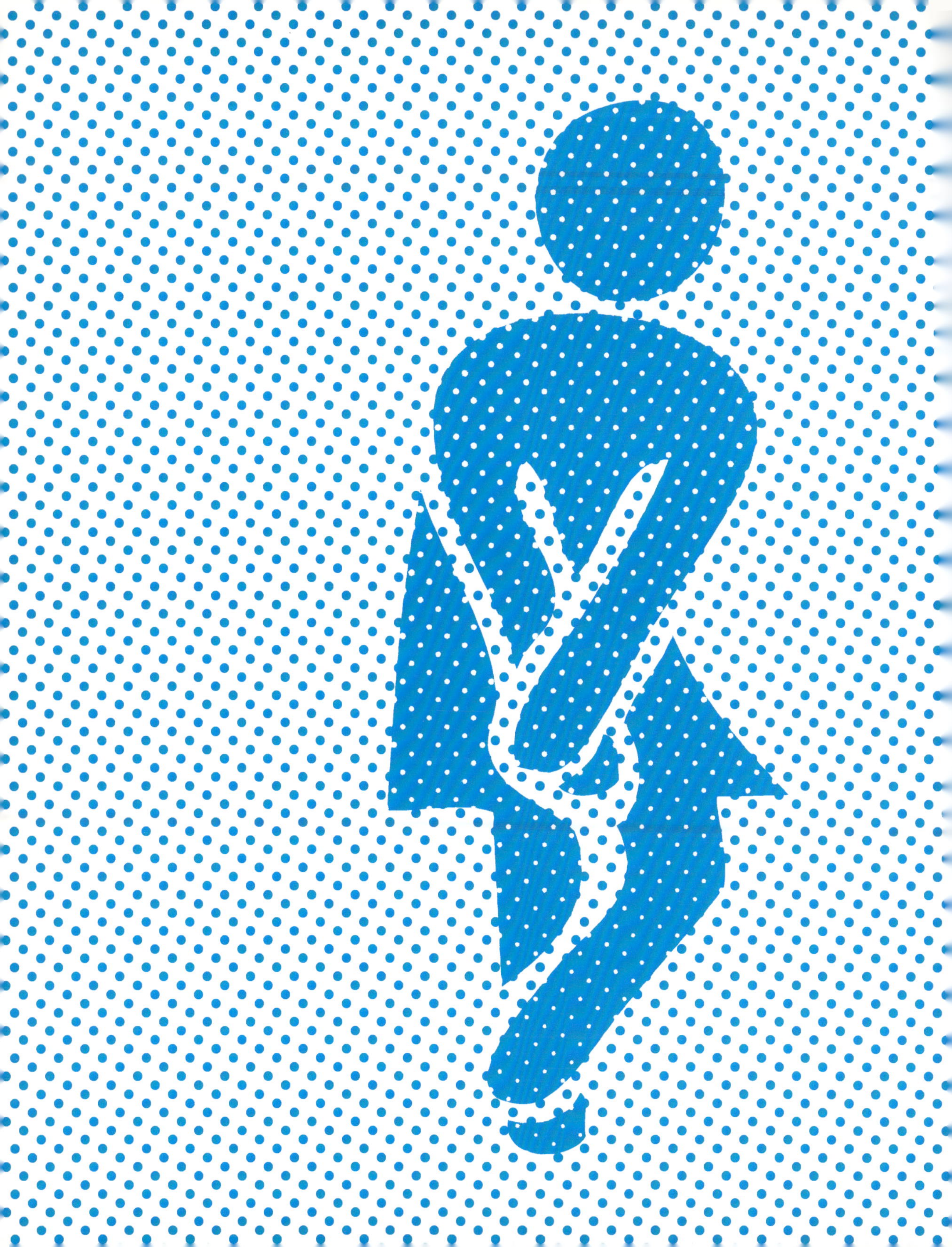

Inhalt
Content

Rein in den Dschungel | Into the jungle 9

Wegweiser | Directions 17

Codierte Schätze | Coded Treasures 21

Uncodierte Schätze | Uncoded Treasures 75

Einzelheiten | Details 95

Raus aus dem Dschungel | Out of the jungle 193

Sprachschatz | Vocabulary 201

Literaturempfehlungen | Further Readings 243

Index | Index 251

Impressum | Imprint 275

Blumhardt Verlag | Publishing House 289

TOILET
FOR
LADIES

Rein in den Dschungel
Into the jungle

Du Tarzan – Ich Jane
Fundstücke aus einer gender codierten Welt

Der Dschungel an Produkten hat uns übermannt: Produkte so typisch Tarzan oder so typisch Jane! Einmal im Dschungel gefangen begann das Abenteuer dieser Objektsammlung. Während des Sammelns der in diesem Buch vorgestellten Produkte, Verpackungen oder Grafiken wurden wir von den eindeutigen stereotypen geschlechterspezifischen Zuordnungen überrascht.

Wir wissen, dass es einen Unterschied zwischen dem männlichen und weiblichen Gehirn und den Verhaltensweisen von Mann und Frau gibt. Doch ist dies der Grund für zwei so gegensätzliche Produktwelten: eine kontrastreiche, dunkle, harte, eckige, glänzende, intelligente, starke, geordnete und praktische? Und eine andere helle, rosafarbene, weiche, runde, matte, leichte, schwache, verspielte und dekorative? Die Gründe für die ausgeprägte Verschiedenartigkeit von Frauen und Männern sind nicht nur angeboren. Unser gesellschaftliches Umfeld prägt uns mehr als uns lieb ist. Wir reagieren, wie es von uns erwartet wird, und möchten in keinem Fall »unmännlich« oder »unweiblich« wirken. Auch wenn sich Eltern zunehmend um die genderneutrale Erziehung bemühen, ist dies nahezu unmöglich. Die »heimlichen« Erzieher wie Werbung, Castingshows oder Produkte basieren auf tradierten Rollenbildern.

Die Geschlechterklischees und damit verbundenes Denken lassen uns oft unfreiwillig in stereotypische Verhaltensweisen schlüpfen. Tests zeigen, dass Mädchen nicht grundsätzlich schlechter in Mathematik oder unpraktischer als Jungen sind und dass genauso Jungen ▸

You Tarzan – Me Jane
Interesting stories from a gender-coded world

We are overwhelmed by the product jungle: Products are either typically Tarzan or typically Jane!
When we got lost in the jungle, the adventure of collecting objects started. While we were compiling products and artwork presented in this book, we actually got caught by surprise coming across these obvious stereotypes of gender-specific classification.

We now know that there is a difference between the male und female brain, and between male and female behaviour. But is this a reason for two diametrically opposite product worlds? A high contrast, dark, hard, edgy, glossy, intelligent, powerful, categorised, practical world and another one described best as being bright, pink, soft, rounded, matte, light, weak, playful and decorative. The reasons for the pronounced differences between men and women are not solely inborn. Our social environment has a greater influence than we would like. We respond in a fashion expected of us and do not wish to be seen as being »unmanly« or »unfeminine«. Even if parents are increasingly trying to bring up their children in a gender-neutral way, this is virtually impossible. The »secret« trainers such as advertising, casting shows and products are all based on traditional gender roles.

The gender stereotypes and associated mindsets often result in an unintentional slide into stereotyped behaviour patterns. Tests show that girls are not basically worse at maths or less practical than boys, and similarly, boys are not emotion-less or less sensitive than girls. Yet, from an early age onwards, subtle comments influence us; ▸

nicht weniger emotionslos oder feinfühlig sind als Mädchen. Von klein auf gibt es jedoch subtile Äußerungen wie: »Du bist doch ein großer Junge und der ist tapfer und stark und weint nicht«.

Viele Unternehmen bedienen sich bei der Gestaltung und Vermarktung von Produkten dieser klischeehaften Stereotype und Gender Codes – und dies mit großem Erfolg. Eine Frau würde heute kaum eine Kosmetikserie kaufen, die sich als typisch männliches Produkt darstellt. Es geht soweit, dass die Verträglichkeit in Frage gestellt wird. Eine Entwicklung, die in den letzten Jahren durch die stärkere Produktdiversifikation beschleunigt wurde.

Unsere Sammlung von Objekten ist subjektiv und plakativ – es sind Fundstücke aus dem Produktdschungel, der uns täglich umgibt. Mit der bewussten Gegenüberstellung der Objekte zeigen wir Aspekte des Gender Design. Eine geschlechtliche Zuordnung der Objekte ist an verschiedenen Merkmalen ablesbar. Neben der Unterscheidung durch das biologische Geschlecht oder stereotype Rollenbilder, sind Farbcodierung, Formsprache, Materialität sowie Komplexität oder Proportionen gestaltungsbestimmend.

Neben den codierten Objekten haben wir in unsere Sammlung auch einige Produkte aufgenommen, die auf Gender Codes und somit eine geschlechtsspezifische Ausrichtung verzichten und bei denen eine Produktdifferenzierung nach Funktion, Ergonomie oder Anwendungspräferenzen stattfindet. ▸

as, »You're a big boy, and big boys are brave and strong and don't cry!«

Many companies take advantage of stereotypes and gender codes when designing and marketing their products – and they do so with great success. It is hardly likely that any woman will buy a product from a cosmetics range, which presents itself as a typical male product. Things have even got to the point where compatibility is put into doubt. A development that has recently been accelerated by increased product diversification.

The collection presented in this book is subjective, demonstrative, and striking – all artefacts were selected within the jungle of products surrounding our daily lives. The deliberate confrontation of items features many aspects of Gender Design. Differentiating along biological gender and stereotyped role models, designs are driven by colour-coding, design language, materiality, and complexity as well as proportions.

Besides the coded objects, our collection also includes some products, which abandon the use of gender codes and any gender-specific assignment, differentiating products by function, ergonomics or user/application preferences.

Complementary to the illustrations of the artefacts, we have also included original product descriptions and press releases by the original manufacturers and distributors in the second part of the book. ▶

Neben den Abbildungen der Fundstücke finden Sie im zweiten Teil des Buches originale Produktbeschreibungen und Pressetexte von Herstellern oder Vertrieb. Einige Objekte werden in ihrem Gestaltungszusammenhang dargestellt – dies sind zum Teil Werbedarstellungen oder Beschreibungen über die Gestaltung der Internetpräsenz. In weiten Teilen wird hier auf die Eigendarstellung der Hersteller Bezug genommen. Dadurch entstehende mögliche Wertungen wurden nicht bereinigt. Bei einigen Produkten erlauben wir uns eigene Interpretationen, die durch die Unterschiedlichkeit der darstellten Objekte provoziert wurden.

Im Glossar des Kapitels *Sprachschatz* finden Sie erklärende Zitate zu Begriffen wie Gender, Gender Codes etc.

Die Sammlung wurde im Rahmen des Forschungsprojektes *Genderspezifische Gestaltung von Produkten und Nutzungskonzepten* von der Hochschule Hannover und der Hochschule für Technik und Wirtschaft Berlin zusammengetragen und möchte auf anschauliche Art und Weise die Kategorie *Gender* in der Gestaltung verdeutlichen. ■

Birgit Weller und Katharina Krämer

Some objects are presented within their design contexts – starting by advertisement, product descriptions, and the design of websites. Many items refer to manufacturers' original presentations. No evaluations/value judgements arising have been altered or removed. In the case of some products we have also added our own interpretations, provoked by the disparity of the objects as shown.

In the *Vocabulary* section we have included a glossary to explain terms such as Gender, Gender Codes, etc.

The collection was compiled as a part of a research project entitled *Gender-specific designer products and usage concepts* by the Hochschule Hannover and the HTW Berlin and seeks to illustrate vividly the category of gender and the classification of *gender* in design. ■

Birgit Weller and Katharina Kramer

MA

Wegweiser
Directions

Wegweiser – Was Sie wo finden

Im ersten Bildteil *Codierte Schätze* finden Sie Abbildungen der codierten Fundstücke. Wir haben sie paarweise gegenübergestellt. Der zweite Bildteil *Uncodierte Schätze* zeigt jeweils ein Objekt je Doppelseite.

Jeweils auf den rechten Abbildungsseiten stehen unter der Seitenangabe Zahlen. Diese verweisen auf die Seiten im Kapitel *Einzelheiten*. Dort finden Sie zu den jeweiligen Objekten Beschreibungen, Pressetexte, Gestaltungszusammenhänge sowie detailliertes Bildmaterial. Zwei Lesezeichen geben die Möglichkeit, die zusammengehörigen Bild- und Detailinformationen zu markieren.

Der Detailteil *Einzelheiten* ist zweisprachig: Die linke Seite beinhaltet die deutschen Texte, die rechte Seite die englischen Übersetzungen. Originale Produktbeschreibungen, Pressetexte sowie Passagen aus der jeweiligen Internetpräsenz sind jeweils kursiv gesetzt und an den Zitatenden mit Fußnoten versehen. Die Nachweise für diese Quellen sowie Informationen zu Produktbenennung, Hersteller und Bildnachweise finden sich im Kapitel *Index*.

Der *Sprachschatz* ist ein Glossar mit erklärenden Zitaten zu themenrelevanten Begriffen.

Eine Liste von Publikationen und Internetseiten zu *Gender* und *Gender im Design* ermöglicht eine weiterführende Auseinandersetzung. ■

Directions – Where you can find what

The first selection of *Coded Treasures* shows pictures of coded artefacts. Each page presents paired items. The second section, *Uncoded Treasures*, shows one object on a double page spread.

The right-hand page of the picture pairs also includes a reference number under the page number. The reference numbers refer to pages in the *Details* chapter. This section includes descriptions, press releases, design context and detailed picture material. The two bookmarks are useful in investigating how pictures and detail information fit together.

The *Detail* section is bilingual: the left-hand side presents the German text, the right-hand page the English translation. Original product descriptions, press releases and passages from the various websites are italized, footnotes are found at the end of the quotations. References for sources and information about product designations, manufacturers and picture credits are listed in the *Index*.

The *Vocabulary* includes a glossary with explanatory notes on relevant terms.

A list of publications and websites about *Gender* and *Gender in Design* is also provided for further research. ■

Codierte Schätze
Coded Treasures

01 Petty & Pink Collection; Esprit
Urban Street Collection; Esprit

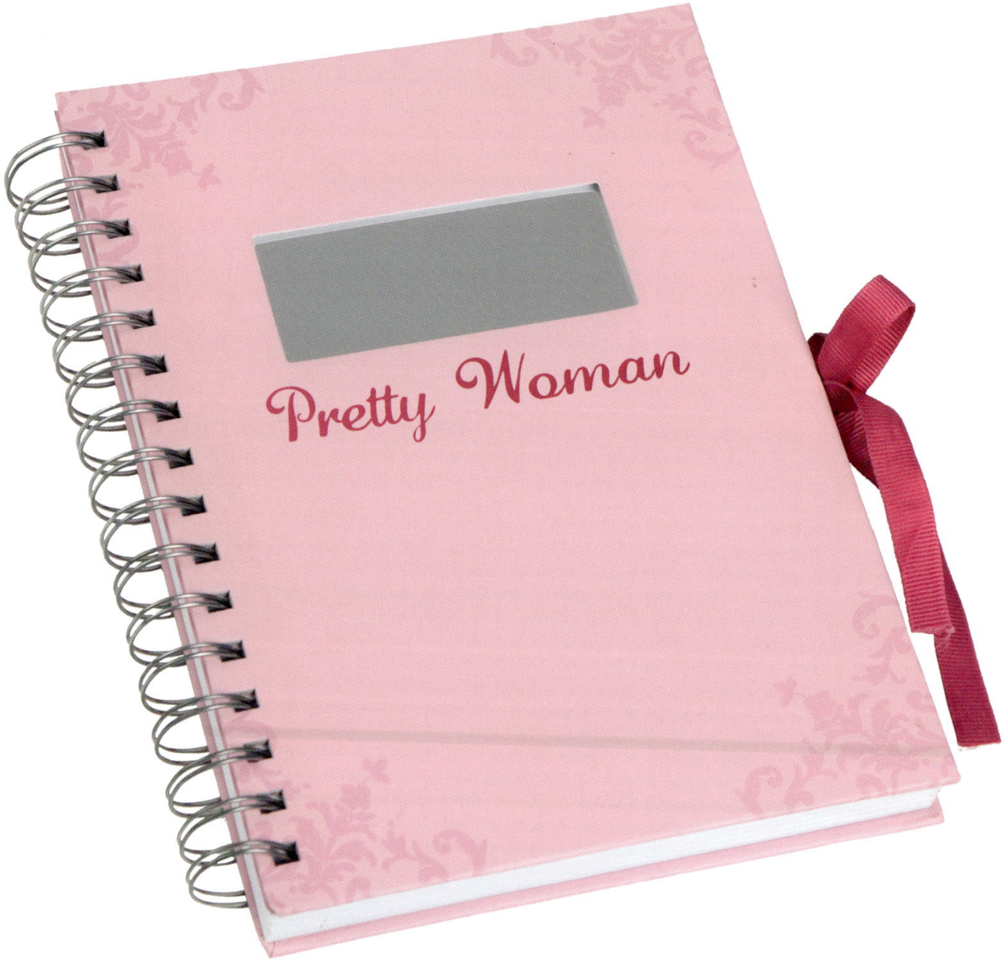

▸ 96

02 Glamour & Style Collection; Esprit
Blue & Blue Collection; Esprit

▶ 100

03 Fuel for Life Woman; Diesel
Fuel for Life Man; Diesel

▸ 102

0,33L
Coca-Cola
Schutzmarken
light

▸ 108

05 Camee Toiletries bag lips; Vlieger & Vandam
Camee Toiletries bag moustache; Vlieger & Vandam

▸ 110

06 EVE; Eve Getränke-GmbH
Holsten Pilsener; Holsten Brauerei

▶ 112

HOLSTEN
HOLSTEN-BRAUEREI AG
SEIT 1879
HOLSTEN
Pilsener
PREMIUM
HAMBURG · GERMANY

07 Anti Aging Bier; Klosterbrauerei Neuzelle
Marathon; Klosterbrauerei Neuzelle

▸ 118

08 B.United Jeans Woman; Benetton
B.United Jeans Man; Benetton

▶ 122

09 Pearl & Beauty, Double Effect; Nivea
Dry Impact, Cool Kick; Nivea

▶ 124

NIVEA
FOR MEN
DRY
IMPACT
24h
REGULIERT
DIE TRANSPIRATION
► MINERALIEN ◄
RÉGULE
LA TRANSPIRATION
► MINÉRAUX ◄
ANTI-TRANSPIRANT
01177010
NIVEA
FOR MEN
24h
EXTRA FRISCHE-KICK
► COOL-CARE FORMEL ◄
ANTI-TRANSPIRANT

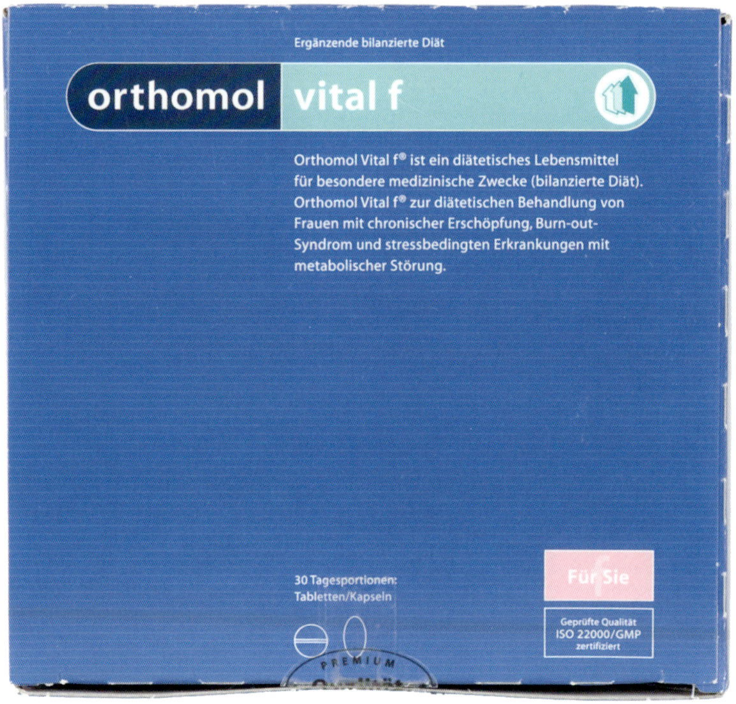

10 Vital f; Orthomol
Vital m; Orthomol

▸ 126

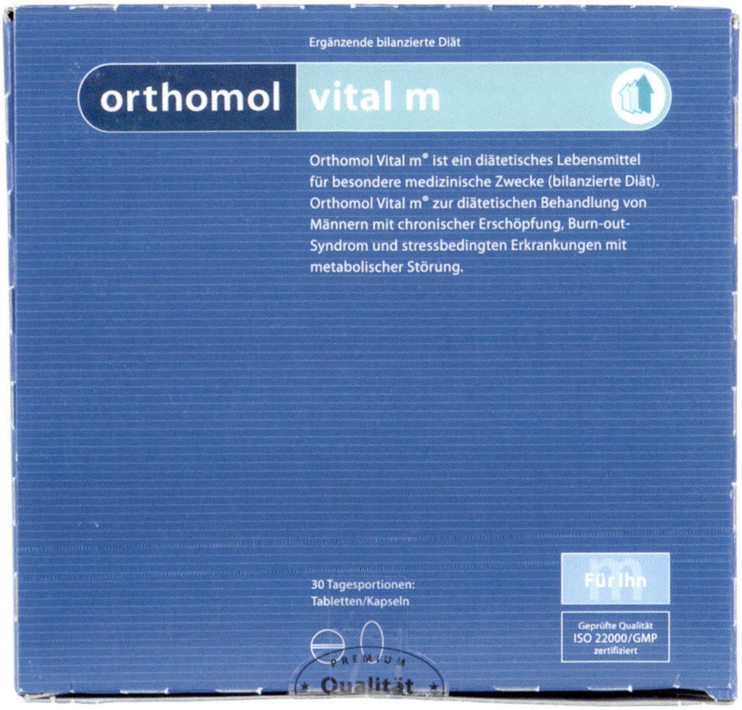
Ergänzende bilanzierte Diät
orthomol vital m
Orthomol Vital m® ist ein diätetisches Lebensmittel für besondere medizinische Zwecke (bilanzierte Diät). Orthomol Vital m® zur diätetischen Behandlung von Männern mit chronischer Erschöpfung, Burn-out-Syndrom und stressbedingten Erkrankungen mit metabolischer Störung.
30 Tagesportionen:
Tabletten/Kapseln
Für Ihn
Geprüfte Qualität
ISO 22000/GMP
zertifiziert

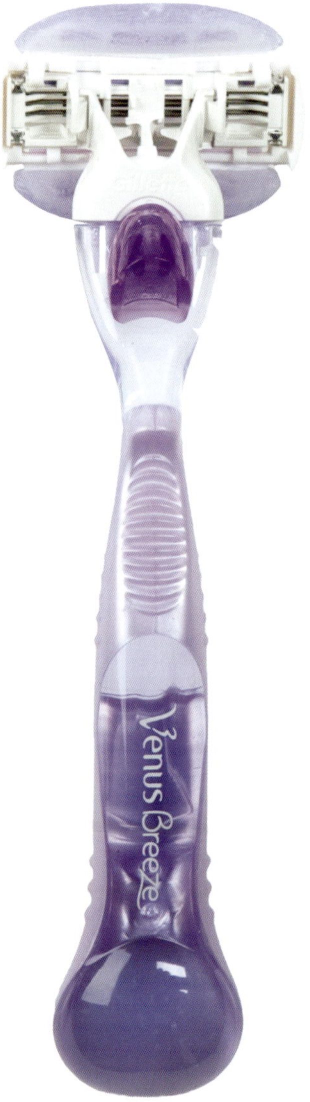

11 Venus Breeze; Gillette
Mach 3; Gillette

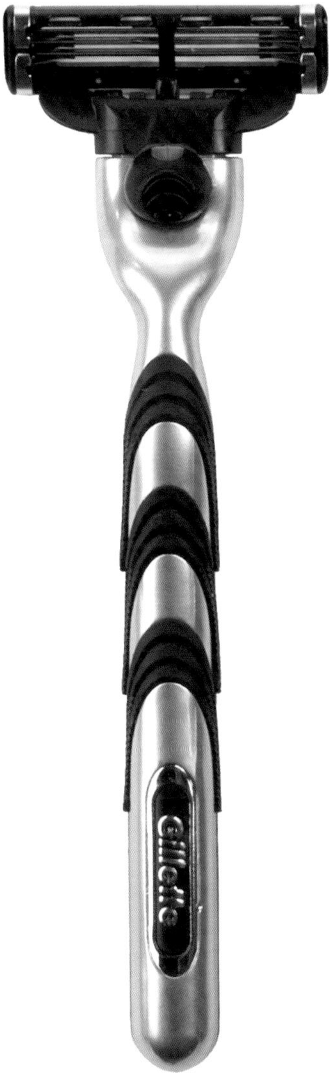

▸ 128

12 Supreme, Go Fresh, Beauty; Dove Extra Fresh, Clean Comfort, Aqua Impact; Dove Men+Care

► 132

NEW
Dove
MEN+CARE
EXTRA FRESH
KÜHLENDE WIRKUNG
PFLEGEDUSCHE FÜR KÖRPER UND GESICHT
GEL DOUCHE HYDRATANT VISAGE ET CORPS
FÜR GEPFLEGTE HAUT
NEW
Dove
MEN+CARE
CLEAN COMFORT
MILDE FORMEL
PFLEGEDUSCHE FÜR KÖRPER UND GESICHT
GEL DOUCHE HYDRATANT VISAGE ET CORPS
FÜR GEPFLEGTE HAUT
NEW
Dove
MEN+CARE
AQUA IMPACT
MEERESMINERALIEN
PFLEGEDUSCHE FÜR KÖRPER UND GESICHT
GEL DOUCHE HYDRATANT VISAGE ET CORPS
MICRO MOISTURE

13 Contradiction; Calvin Klein
Contradiction for men; Calvin Klein

▸ 136

14 Telephone Plush pink; Silly Gifts, Present Time
Telefon T3, Jacob Jensen

▸ 138

15 LEGO® Bauspaß für Mädchen
LEGO® Technic

▸ 140

OGEL

▸ 142

17 Mutter; Depesche Vertrieb
Papa; Depesche Vertrieb

► 144

Du bist in jedem
Alter der beste
Papa
Alles Liebe zum
neuen
besten Alter!

18 Bosch private collection; Theo Klein
Bosch mini; Theo Klein

▸ 146

19 Pliers Combo (Pink); Tomboy Tools
Zangen; Wiha

▸ 150

20 Vileda junior; Theo Klein
Bosch mini; Theo Klein

▸ 154

21 Classic Alox; Victorinox
Handyman; Victorinox

▸ 158

22 Air Vignon; menu
Leverpull LM-200; Screwpull

▸ 160

23 Storchschere; Wasa Solingen
Arbeitsschere; Forum

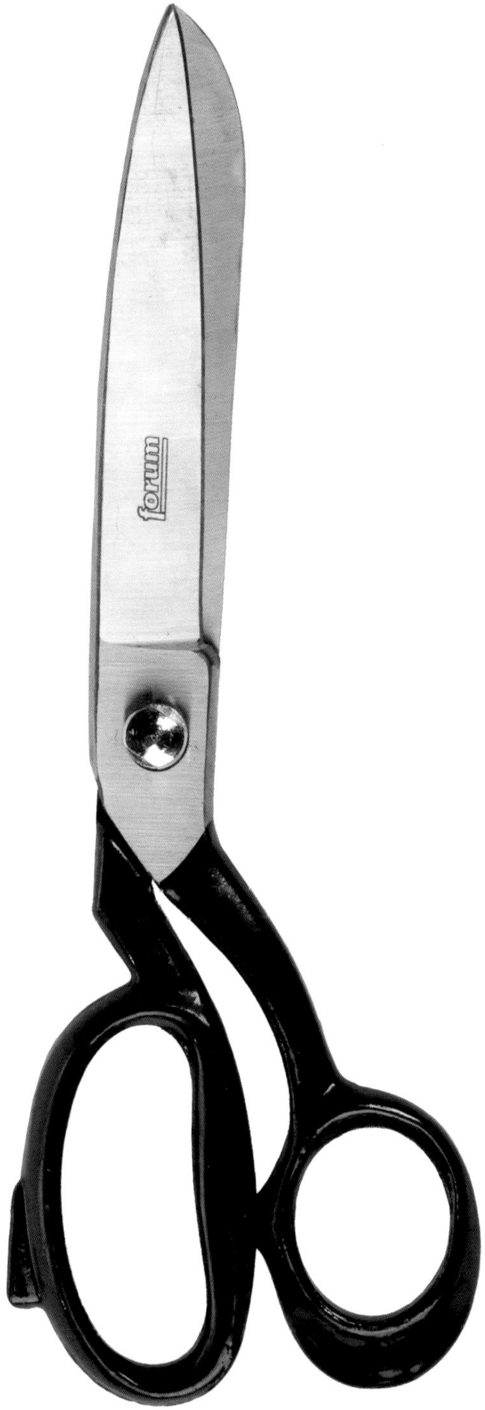

▸ 164

24 Pretty Useful Tools™
Hammer; Picard

▸ 166

25 sixx; fem Media
DMAX; Discovery Communications Deutschland

▶ 170

DIE MEISTE KOHLE HAUEN SIE AUF DER ARBEIT RAUS.
HELDEN UNTER TAGE
SCHICHTBEGINN FREITAGS 22:15 UHR AUF DMAX
NEUE SERIE
DMAX
DMAX.DE

26 Kadett 912 Ideal Westernland; J.G.Schrödel
VIP Ideal Westernland; J.G.Schrödel

▸ 172

Uncodierte Schätze
Uncoded Treasures

27 Ergomat; Fiskars

77

▶ 174

28 2911 Softouch®, Fiskars

▶ 176

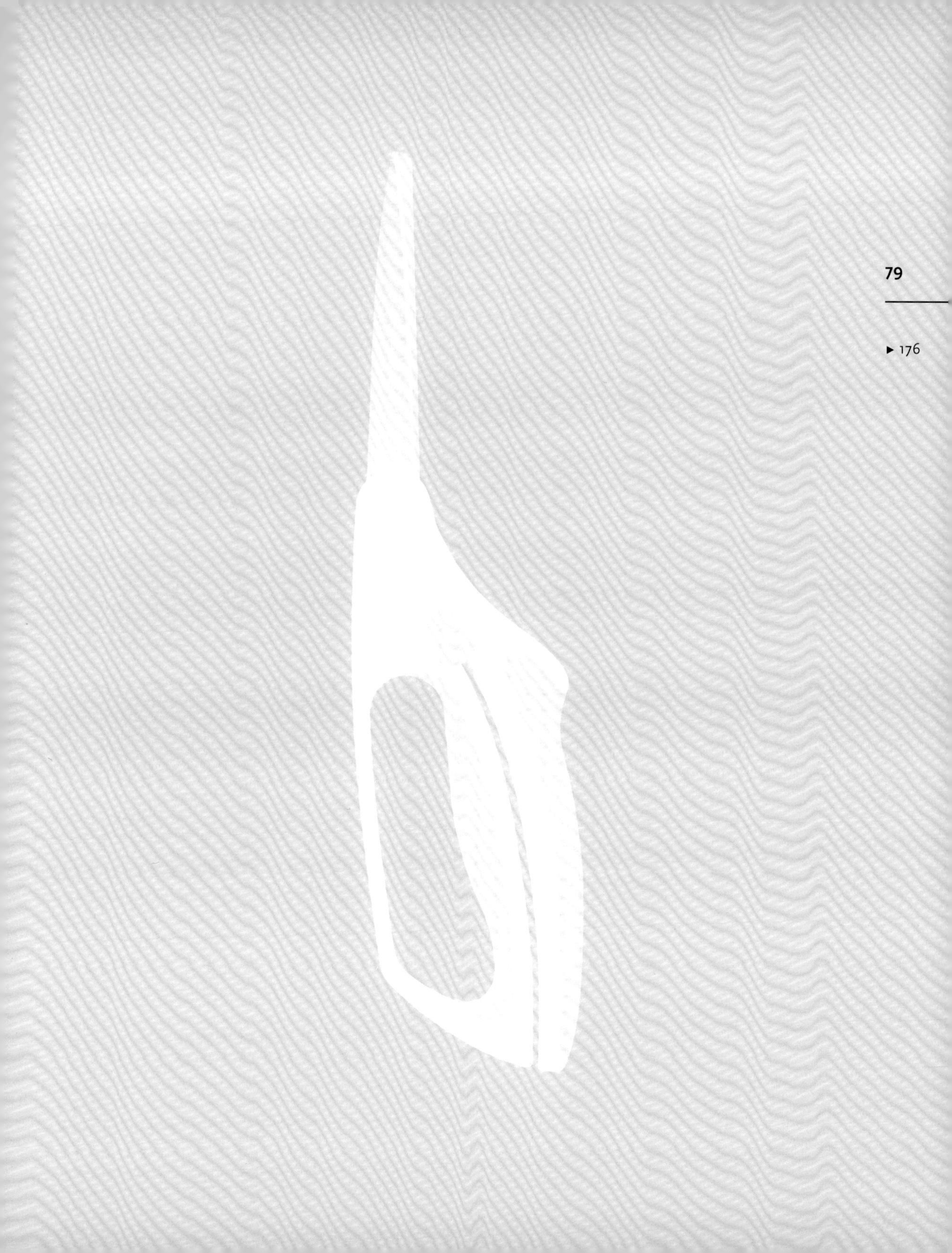

29 IXO; Bosch

▸ 178

BAHCO

▶ 180

▶ 182

32 Cookware, Royal vkb

▶ 184

33 brand eins; brand eins Verlag

▸ 186

brand eins
Wirtschaftsmagazin
Respekt
Es ist Zeit.
Schwerpunkt Respekt

34 Rollbag; Authentics

▶ 188

35 Funnel; Normann Copenhagen

▶ 190

Einzelheiten
Details

01 Petty & Pink Collection; Esprit
Urban Street Collection; Esprit

◂ 22

01 Detail Notizbücher Esprit
01 Detail Notebooks Esprit
02 Notizbuch ›Pretty Woman‹
02 Notebook ›Pretty Woman‹

In der Accessoirelinie von *Esprit* werden die jugendlichen Adressaten gezielt in ihrem gesellschaftlich geprägten Gender-Bewusstsein angesprochen.

Die Produkte für Mädchen beleuchten eine auf Äußerlichkeiten konzentrierte Welt. Mit floralen Ornamenten und Spiegelflächen versehene Notizbücher zeigen die Käuferin selbst als *Pretty Woman*. ▸

01

02

In *Esprit's* accessory product line, the young clientele is deliberately targeted by way of their socially imprinted gender awareness.

The products for girls highlight a world focused on superficial looks. The notebooks with their floral ornamentation and mirrored surfaces show the female buyers themselves as the *Pretty Woman*. Bling decorated plasters awaken the desire to glamorously embellish one's own body, and bring the female buyers into early contact with ›luxury‹. Because of their superficiality and their cheap appearance, the products are devoid of any sense of day-to-day practicality, and therefore also diminish the value of their female users. ▸

Strassverzierte Pflaster wecken den Reiz, den eigenen Körper glamourös in Szene zu setzen und zeigen den Käuferinnen frühzeitig, was ›Luxus‹ ist. Die Produkte verlieren durch ihre oberflächliche und billige Erscheinung ihre Alltagstauglichkeit und werten somit auch die Nutzerinnen ab.

Ein Notizbuch mit *Einband in blauem Design und Patch, Stitchings und Prints, ideal für Schule, Spaß, Büro und Home* bildet das männliche Produktpendant. Hier zeigen die Gestaltungselemente, dass das Leben von Jungen sich aktiv auf der Straße abspielt – mit Jeans, Aufnähern, Musik und Autos. ■

03

04

The matching male product is a notebook with a cover *with a blue design and patch, stitchings and prints, ideal for school, fun, office and home.* The design elements here show that boys live their lives actively on the street – with jeans, patches, music and cars. ■

03 Detail ›Urban Street Collection‹
03 Detail ›Urban Street Collection‹
04 Pflaster mit Swarovski-Steinen
04 plasters with Swarovski bling

02 Glamour & Style Collection; Esprit
Blue & Blue Collection; Esprit

◂ 24

05

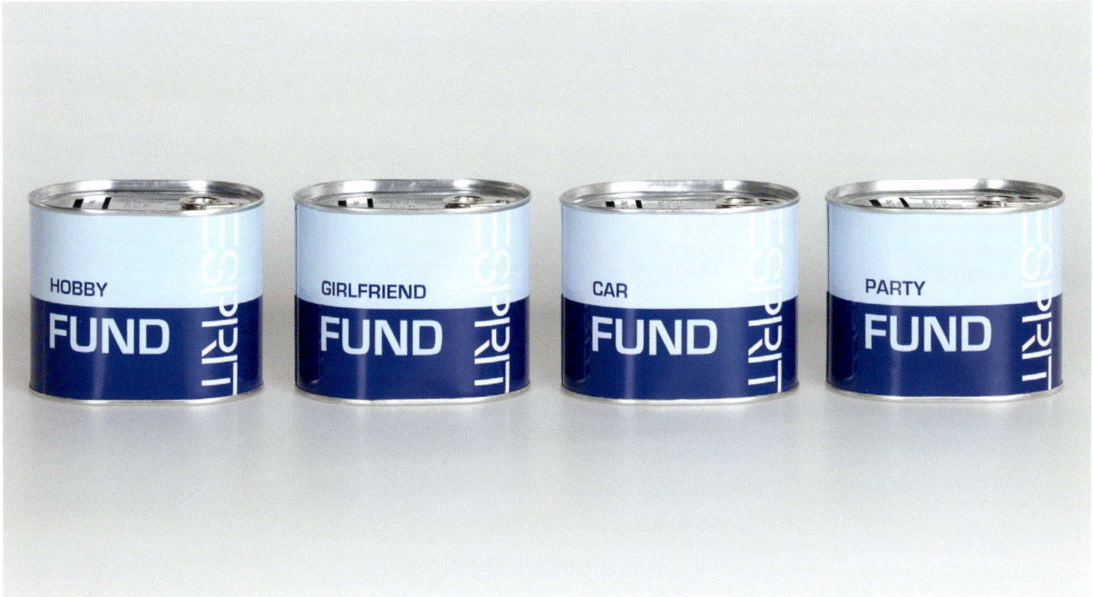

05 Spardosen für Hobby, Girlfriend, Car und Party
05 money boxes for hobby, girlfriend, car and party
06 Spardosen für Dessous, Schmuck, Handtaschen und Nachtleben
06 money boxes for lingerie, jewellery, handbags and nightlife
07 Detail Spardose
07 Detail Money box

Die bis auf die Farbgebung formal identisch gestalteten *Glamour & Style Collection* Spardosen haben eine klare Botschaft: Mädchen sparen für Dessous, Handtaschen, Urlaub, Schmuck, Nachtleben und Schuhe – Jungen sammeln ihr Geld für Party, Auto, Studium, Freundin, Hobby und Urlaub. ■

Formally identical in design except for their colour, the *Glamour & Style Collection* money boxes send out a clear message: girls save for lingerie, handbags, holiday, jewellery, nightlife and shoes – boys save their money for party, car, college, girlfriend, hobby and holidays. ■

07

◂ 26

03 Fuel for Life Woman; Diesel
Fuel for Life Man; Diesel

08

Diesel nutzt die Verpackung für die Geschlechtszuordnung. Der formal identische Flakon unterscheidet sich beim Damen- und Herrenduft durch Details: Golddruckbuchstaben und ein glatter, glänzender Verschluss in Gelbgold beim weiblichen Flakon, welcher zusätzlich von einer Art Netzstrumpfhose umhüllt wird. Dieses ›Kleid‹ kann nur über den Flaschenhals ›ausgezogen‹ werden.

Der männliche Flakon mit dem transparent geprägten Schriftzug besitzt einen strukturierten, mattierten Verschluss in Rotgold und wird von einer Hülle aus grobem Textil umschlossen. Ein robuster Metallreißverschluss öffnet die Hülle, der Flakon wird ausgezogen und weckt so als männlich konnotiertes Attribut die Assoziation einer ›Männerhose‹. ▸

Diesel uses the packaging to differentiate the sexes. The details of the formally identical flacon differentiate the male and female fragrance: printed gold letters and a smooth glittering top in yellow gold on the feminine flacon, which is also surrounded by a type of net stocking. This ›dress‹ can only be ›taken off‹ via the neck of the bottle.

The male flacon with the transparent engraved lettering has a structured, mat top in red gold and is surrounded by a covering of coarse textile material. A rugged metal zip opens the covering, and the flacon is pulled out – an attribute with a direct masculine connotation, creating an association with a pair of ›men's trousers‹.

Diesel offers a range of different ›clothes‹ for the two fragrances – the analogy here is with a clothes cupboard, to select the fitting outfit for the fragrance. ▸

08 Hüllenvarianten für ›Fuel for Life Woman‹ und ›Fuel for Life Man‹

08 Cover alternatives for ›Fuel for Life Woman‹ and ›Fuel for Life Man‹

09 Detail ›Fuel for Life Man‹

09 Detail ›Fuel for Life Man‹

09

Für beide Düfte bietet *Diesel* verschiedene Variationen der Umhüllung an – ähnlich wie im Kleiderschrank kann hier das passende Outfit für den Duft gewählt werden.

Bei der Bewerbung der Duftlinie für Frauen *Fuel for Life Unlimited* auf der Website *www.diesel-parfums.de* wird ein sehr ausführliches Frauenbild gezeichnet und mit der Gestaltung des Flakons in Beziehung gesetzt: *Was für eine Frau sind Sie? Sind Sie bereit, grenzenlos zu leben? Bereit, das Leben aufzusaugen? Den Moment zu genießen ohne Angst vor Maßlosigkeit? Willkommen in der Welt von Fuel for Life Unlimited, dem neuen Lebenselixier, nur für Frauen. Ein wertvoller und raffinierter Zaubertrank, reserviert für rebellische Frauen auf der ewigen Jagd nach mehr Vergnügen, Energie und Sinnlichkeit. […] Fuel for life Unlimited ist ein zeitloser Duft, eine dramatische Explosion der Extreme und der Epochen, angefangen von der sorglosen Verrücktheit der »Roaring Twenties« bis hin zu den anmaßenden und extravaganten Vergnügen von Heute. Dieses neue goldene Zeitalter erweckt Federboas, Netzstrumpfhosen, Perlenketten, Kosmetikspiegel und High Heels wieder zum Leben, um die Weiblichkeit zu übertreiben und Genuss neu zu definieren.* ▸

10

10 Hüllenvariante für ›Fuel for Life Woman‹
10 Cover alternatives for ›Fuel for Life Woman‹
11 Detail Deckel ›Fuel for Life Woman‹
11 Detail cap ›Fuel for Life Woman‹
12 Detail Deckel ›Fuel for Life Man‹
12 Detail cap ›Fuel for Life Man‹

11

12

The advertising for the *Fuel for life Unlimited* perfume product line for women on the www.diesel-parfums.de website paints a very detailed picture of women and places this in relationship to the design of the flacon: *What type of woman are you? Are you willing to live life without limits? Willing to completely embrace life? Enjoy the moment with no fear of overstepping the limits? Welcome to the world of Fuel for Life Unlimited, the new elixir of life, only for women. A valuable and sophisticated magic potion reserved for rebellious women permanently on the hunt for more pleasure, energy and sensuousness [...]* ▸

13 Verpackung Duftset ›Fuel for Life Man‹
13 Packaging perfume set ›Fuel for Life Man‹
14 Verpackung Duftset ›Fuel for Life Woman‹
14 Packaging perfume set ›Fuel for Life Woman‹

13

Ein schwindelerregender Aufstieg zum Gipfel grenzenloser Weiblichkeit – ewig glitzernd und sinnlich – der mit Komplizinnen gelebt und geteilt werden muss. Eine grenzenlose Erfahrung … nur für Frauen.

Was, wenn Ihr Parfum-Flakon sich mit der Sinnlichkeit der kühnsten Frauen messen könnte? Der Frau, die zugleich geheimnisvoll und faszinierend, exzessiv und edel ist? Ein fesselnder Körper, graviert mit goldenen Buchstaben, ist eingehüllt in ein Geheimnis aus Intimität, umfangen von einem filigranen Netz aus braun-goldenen Fäden, verziert mit einer eleganten goldenen Kette […] Entdecken Sie die wertvolle Phiole, die die Geheimnisse einer emanzipierten, kühnen und dynamischen Frau unterstreicht.

Fuel for Life Unlimited tritt energiegeladen, frisch und außergewöhnlich sinnlich auf – eine Duft-Erklärung für alle Frauen, die der Nüchternheit den Rausch vorziehen, die Gegenwart der Zukunft und die Kühnheit der Unterwerfung. [03.1 03.2] ■

14

Fuel for Life Unlimited is a timeless fragrance, a dramatic explosion of extremes and the epochs, starting with the careless craziness of the ›Roaring Twenties‹ all the way up to the pretentious and extravagant pleasures of today. This new golden age is conjured up by feather boas, net stockings, pearl necklaces, make-up mirrors and high heels – brought to life to exaggerate femininity and redefine enjoyment. A dizzying ascent to the summit of boundless femininity – eternally sparkling and sensuous – that has to be experienced and shared with fellow conspirators. A limitless experience … only for women.

What if your perfume bottle can match the sensuality of the most daring women? Women who are both secretive and fascinating, excessive and aristocratic?

A captivating body, engraved with golden letters, shrouded by a mysterious intimacy, draped in a delicate network of golden-brown threads, decorated with an elegant gold chain […] Discover the valuable phials, which accentuate the secrets of an emancipated, daring and dynamic woman.

Fuel for Life Unlimited has an energy-laden, fresh and unusually sensuous presence – a perfume statement for all women who prefer intoxication to sobriety, the moment to the future, and daring to subjugation. [03.1, 03.2] ■

04 Coca-Cola light Coca-Cola zero

◂ 28 *Coca-Cola light* und *zero* setzen auf die Geschlechtertrennung. Es werden unterschiedliche Getränke für die unterschiedlichen Bedürfnisse von jungen Frauen und Männern angeboten. Wie diese Bedürfnisse aussehen, beschreibt der Konzern in seinen Pressemitteilungen:

Coca-Cola light – »nur du bist du«
Mit diesem Claim stellt Coca-Cola light die Individualität und Einzigartigkeit junger Frauen in den Mittelpunkt. Er steht für den Aufruf an alle Frauen, sich einfach mal so zu verhalten, wie sie sind, jede auf ihre individuelle Art. Nicht Konformismus ist gefragt, sondern die Fähigkeit, spontan und ungezwungen »mal anders« zu sein – verspielt, laut, lustig, eben so, wie man sich gerade fühlt.[04.1]

Coke Zero Game
[...] Zur Story: Tim, der Star aus dem Coke Zero Werbespot, hat keinen guten Tag erwischt: Erst verlässt ihn seine Freundin, dann sind alle Tickets für das Finale der UEFA EURO 2008™ ausverkauft und zu guter Letzt geht auch noch sein Fernseher kaputt, als er das Finale mit seinen Kumpels schauen möchte. Das Blatt wendet sich für Tim als er mit seinem letzten Geld eine Coke Zero aus einem Automaten zieht und den ersten Schluck aus der Flasche nimmt. Vor ihm bleiben zwei Sportwagen mit quietschenden Reifen stehen und seine blonde Göttin aus dem TV-Spot in Begleitung von zwei heißen Coke Zero Girls laden ihn auf eine actionreiche Reise ein, um Fußball doch noch so zu erleben, wie er sein soll: im Stadion, mit seinen besten Kumpels, umgeben von heißen Girls. [...] [04.2]

»light meets zero« Im Juli wird eine nationale Handelspromotion gemeinsam mit Coke Zero die Verkaufsförderung am Point of Sale unterstützen. Unter dem Motto ›light meets zero‹ können die Konsumenten je zwei spezielle Beachtücher zum Aneinanderknüpfen erwerben, die in schwarzem und weißem Design auf die unterschiedlichen Zielgruppen der Produkte hinweisen – light (weiß) für Frauen, zero (schwarz) für Männer. [04.3] ■

15

15 Detail Getränkedose ›Coca-Cola zero‹
15 Detail beverage can ›Coca-Cola zero‹
16 Detail Getränkedose ›Coca-Cola light‹
16 Detail beverage can ›Coca-Cola light‹

16

Coca-Cola light and *zero* emphasise the differences between the sexes. Different drinks are offered to satisfy the different needs of young men and women. The company describes the nature of these needs in its press releases:

Coca-Cola light – »only you are really you«
This is the claim with which Coca-Cola light places the individuality and uniqueness of young women into the spotlight. It is a call to all women to simply behave the way they really are – everyone in their own individual way. Instead of conformism, the desirable attributes are the ability to »be different sometimes« spontaneously and freely – to be playful, loud, cheerful – precisely the way you feel at that moment.[04.1]

Coke Zero Game
[...] The story: Tim, the star from the Coke Zero commercial, is having a bad day: first his girlfriend leaves him, then all the tickets for the UEFA EURO 2008™ final are sold out, and last but not least, the television breaks down when he wants to watch the final with his mates. Things take a turn for the better for Tim though when he uses his last bit of loose change to get a Coke Zero from a drinks dispenser, and takes his first sip from the bottle. Two sports cars with squealing tires suddenly stop in front of him, and his blond goddess from the TV commercial accompanied by two hot Coke Zero girls invite him to join them on an action-packed trip so that he finally manages to watch football the way it should be watched: in the stadium, with his best mates, surrounded by hot girls. [...] [04.2]

»light meets zero«
A national retail promotion together with Coke Zero is launched in July to support turnover at point of sale. Under the »light meets zero« slogan, consumers can acquire the two special beach towels which button together and whose black-and-white designs highlight the different target groups of the two products – light (white) for women, zero (black) for men.[04.3] ■

◂ 30

05 Camee Toiletries bag lips; Vlieger&Vandam
Camee Toiletries bag moustache; Vlieger&Vandam

Die Kulturtaschen *Camee Toiletries bag lips* und *Camee Toiletries bag moustache* der niederländischen Designer *Vlieger & Vandam* spielen auf ironische Weise mit der Geschlechterdualität. Stilisierte rollentypische Merkmale von Mann und Frau werden als einziges Unterscheidungsmerkmal für die ansonsten identisch gestalteten Objekte eingesetzt. Die äußerst hochwertige Verarbeitung und die Verwendung edler Materialien geben den Taschen einen seriösen und exklusiven Charakter und unterstreichen die Paradoxie der Kategorisierung von Gender in der Gestaltung. ■

17

The toiletries bags *Camee Toiletries bag lips* and *Camee Toiletries bag moustache* from the Dutch designer *Vlieger & Vandam*, play ironically with the difference between the sexes. Stylised typical differentiating features between men and women are used as the only identifiers of the different sexes in the otherwise identically designed articles.

The extremely high quality craftsmanship and the use of luxurious materials lend the bags a serious and exclusive character, and highlight the paradoxes involved in categorising gender in design. ■

18

17 Produktdetail Geschlechtsspezifikation
17 Product detail gender specification
18 Produktdetail Verarbeitung
18 Product detail finish

06 EVE; Eve Getränke-GmbH Holsten Pilsener; Holsten Brauerei

EVE wird als *leicht alkoholhaltiges Apéritifgetränk für die Frau* bereits seit 2006 in der Schweiz vermarktet. Das Produkt eröffnet ein neues Getränkesegment für Frauen und wird über die Gastronomie und im Handel sowie in Beauty- und Coiffeursalons vertrieben.

◂ 32

Zielgruppe sind *junge, selbstbewusste, moderne, trendbewusste* Frauen zwischen 20 und 40 Jahren, die *nebst dem beruflichen Erfolg stilvollen Spaß im Ausgang haben wollen* und von denen laut Unternehmensumfragen ein Großteil selten oder nie Bier trinkt. *EVE* wird als eine *genussvolle Alternative zu Cocktails und anderen alkoholhaltigen Getränken* beschrieben.

Die Schweizer Biermarke Cardinal zeigt sich in diesem Herbst von ihrer weiblichen Seite und bringt Ende September das neue »EVE« auf den Markt. »EVE« definiert eine neue Kategorie von Getränken und wurde speziell von Frauen für Frauen entwickelt. [...]

»EVE« wird aus Reis und Malz gebraut und ist dadurch prickelnd mild und der naturreine Litchisaft verleiht dem »Apérit-EVE« sein natürlich fruchtiges Aroma. Ob an der Bar, im Club oder beim Apéro unter Freundinnen: »EVE« von Cardinal mit nur 3,1 Volumenprozenten Alkohol ist das perfekte Erfrischungsgetränk nach einem langen Arbeitstag. Im Gegensatz zu Bier wird »EVE« denn auch niemals direkt aus der Flasche getrunken, sondern pur aus dem Eisschrank, im Champagner- oder Longdrinkglas mit Eis serviert. [06.1] ▸

19

19 Detail ›Eve‹
19 Detail ›Eve‹
20 Detail ›Holsten‹
20 Detail ›Holsten‹

EVE has been marketed in Switzerland since 2006 as a *low-alcohol aperitif for women*. The product opens up a new drinks segment for women, and is sold via restaurants, retailers, hairdressers and beauty parlours.

The product is targeted at *young, competent, modern, fashion-conscious* women between 20 and 40, *who in addition to professional success, also want to have stylish fun when they go out, and according to market surveys,* a large proportion of whom rarely or never drink beer. *EVE* is described as a *pleasurable alternative to cocktails and other alcoholic drinks.*

This autumn, the Swiss Cardinal beer brand shows its feminine side and launches the new »EVE« onto the market at the end of September. »EVE« defines a new category of drinks and was specially developed by women for women. [...]

»EVE« is brewed from rice and malt to give it a prickly mild flavour. The pure and natural litchi juice gives the »EVE aperitif« a natural fruity aroma. For bar, club or a drinks reception with friends: »EVE« from Cardinal with only 3.1 volume per cent alcohol is the perfect refreshing drink after a long working day. Unlike beer, »EVE« is never drunk direct from the bottle, but pure from the fridge, served with ice in Champagne or long-drink glasses. [06.1] ▸

20

21

21 Produktpräsentation ›Eve‹

21 Product presentation ›Eve‹

22 Werbedarstellung ›Holsten‹

22 Advertisement ›Holsten‹

Es gibt verschiedene Werbekampagnen – wie das Offerieren des Getränkes *in rund 800 Beauty- und Coiffeursalons sowie Fashion-Boutiquen für stilbewusste Frauen beim Shopping* oder das Promoten des Getränkes durch die Gewinnerinnen des *EVE´s Angels-Casting* – die eine bestimmte Frauenrolle definieren. Es wird das Bild einer auf Äußerlichkeiten und Konsum bedachten Frau gezeichnet, die durch den Genuss von *EVE* erfolgreich und stilvoll wirkt. Unterstrichen wird dieses Bild durch die Bewerbung des Produkts durch die ehemalige Miss Schweiz und *aufstrebende DJane* Mahara Mc Kay: *Nicht nur geschmacklich sind Mahara und EVE auf einer Linie: Auch der Stil verbindet sie.* [06.2]

Auch die Vermarktung in der *sehr feminin anmutenden* 0,25 l Glasflasche, die auch im 4er Pack zur Verfügung steht, impliziert, dass ein herkömmliches 6er Pack für Frauen untragbar – im Sinne des Wortes – ist. Die weich gewölbten Formen sowie die geringe Größe der Flasche unterstreichen den femininen Eindruck. Das klare Glas der Flasche und die zurückhaltende und pastellfarbene Gestaltung der Etiketten zeigen das ebenfalls leicht gefärbte Getränk und vermitteln zusätzliche Leichtigkeit. ▸

A range of advertising campaigns – such as offering the drink *in around 800 beauty parlours and hairdressing salons, or in fashion boutiques for style-conscious women while they are doing their shopping*, or having the drink promoted by the winners of *EVE's Angels-Casting* – define a specific female role. The picture painted here is that of women who care about their looks and like to consume, women who appear successful and stylish by enjoying *EVE*. This image is highlighted by product advertisements featuring the former Miss Switzerland and *up-and-coming DJane* Mahara Mc Kay: *Mahara and EVE not only toe the same line in terms of taste – they also share the same style.* [06.2]

The marketing of the *very feminine looking* 0.25 l glass bottle, which is also available in a 4-pack, also implies that a standard 6-pack is unbearable for women – in the original sense of the word. The gently curved shape and the small size of the bottle emphasise the feminine impression. The clear glass of the bottle and the restrained and pastel-coloured design of the labels also reveal the slight coloration of the drink as well as creating a sense of lightness. ▸

22

Wir Männer freuen uns darauf, endlich mal wieder so sein zu können, wie wir wirklich sind. Darauf stoßen wir mit einem Bier an, das ebenso unkompliziert ist: Holsten Pilsener. Ein bodenständiges, ehrliches und authentisches Bier, das genauso ist, wie wir es sind. Mit Holsten Pilsener können Männer einfach sie selbst sein: sympathisch, echt und geradeaus. Erstklassige Qualität, die man am liebsten gemeinsam mit seinen Freunden genießt. Holsten. Auf uns, Männer! [06.3]

Die Marketingstrategie von *Holsten Pilsener* beschreibt das Bier als reines Männerprodukt. Inhalt und Gestaltung der Website sowie die Werbung gründen auf die klischeehaften Männerdomänen Sport, Grillen und Bier trinken.

Das Layout der Website zeigt handschriftliche Typo auf Bierdeckeln und präsentiert Flaschenöffnersounds. Es werden Merchandising-Produkte wie Biergläser, Spielkarten, Schlüsselbänder, Schreibsets von Parker, Gürtel mit großer Schnalle und T-Shirts angeboten, die das stereotype Männerbild unterstützen. ■

23 Details ›Eve‹ und ›Holsten‹
23 Details ›Eve‹ and ›Holsten‹

We men look forward to finally being able to behave the way we really are. We toast the return to our true selves with a beer that is just as uncomplicated: Holsten Pilsener. A solid, honest and authentic beer that exactly matches the way we feel. Holsten Pilsener allows men to be their true selves: sympathetic, real and straightforward. First class qualities, which one likes to enjoy best in the company of friends. Holsten. Here's to us men! 06.3

The *Holsten Pilsener* marketing strategy describes beer as a pure male product. Content and design of the website, as well as the advertising, are based on clichéd male domains: sport, barbecuing and beer drinking.

The layout of the website shows hand-written lettering on beer mats, and presents the sounds of bottles being opened. Merchandising products such as beer glasses, playing cards, key rings, Parker pen sets, belts with large buckles, and T-shirts are offered, and underpin a stereotyped male image. ■

07 Anti Aging Bier; Klosterbrauerei Neuzelle
Marathon; Klosterbrauerei Neuzelle

◂ 34

24 Details ›Anti Aging Bier‹
24 Details ›Anti Aging Bier‹
25 Details ›Marathon‹
25 Details ›Marathon‹

Anti Aging Bier, das Bier der Schönheit und Vitalität. Unser Anti Aging Bier soll einen kleinen Beitrag zu Ihrem Wohlbefinden leisten. Die Uhr des Lebens lässt sich damit nicht anhalten. Die zugesetzten natürlichen Stoffe Kurbad-Sole, Spirulina (Algen) und Flavonoide (Quercetin) erhöhen zum Schutz der Zellen vor den schädigenden Angriffen freier Radikaler den Anteil an wertvollen Vitaminen, Mineralstoffen und Aminosäuren. Ihrer Gesundheit zuliebe verwenden Sie dieses Bier sorgfältig, es enthält Alkohol. Vermeiden Sie das Anti Aging Bier als Verkehrsteilnehmer, Schwangere, Kranker und Jugendlicher. Das Anti Aging Bier ist im Rahmen eines von der Klosterbrauerei Neuzelle durchgeführten Forschungsvorhabens in Zusammenarbeit mit den Firmen Fermtec Berlin und IGV GmbH (Institut für Getreideverarbeitung) Potsdam entwickelt worden. Hauptbestandteil des Anti Aging Bieres ist das Original Badebier Neuzeller Kloster-Bräu, dem ein Konzentrat aus Sole aus Bad Saarow, Spirulina der Firma IGV aus Potsdam und Flavonoide der Firma Matec aus Wildau zugesetzt wird. Das Anti Aging Bier enthält nur natürliche Stoffe, keine chemischen Zusätze. [07.1] ▸

Anti Aging Beer – the beauty and vitality beer. Our Anti Aging Beer makes a small contribution to improving your sense of well-being. This does not stop the clock though. The added natural ingredients – spa brine, Spirulina (algae) and flavonoids (quercetin) raise the proportion of valuable vitamins, minerals and amino acids to protect the cells from being attacked and damaged by free radicals. For the sake of your health, use this beer prudently, it contains alcohol. Do not drink Anti Aging Beer if you intend to drive. It should also be avoided by pregnant women, people who are ill, and youngsters. The Anti Aging Beer was developed by Klosterbrauerei Neuzelle as part of a research project undertaken in co-operation with Fermtec Berlin and IGV GmbH (Institute for Cereal Processing), Potsdam. The main ingredient of the Anti Aging Beer is the original bath-beer Neuzeller Kloster-Bräu, to which a concentrate has been added containing brine from Bad Saarow, Spirulina from IGV in Potsdam, and flavonoids from Matec in Wildau. The Anti Aging Beer contains only natural ingredients, no chemical additives. 07.1 ▶

24

25

Marathon, das Bier danach. Marathon ist nicht nur ein Genussmittel; es ist mehr. Marathon ist ein Pils, dem L-Carnitin zugesetzt wurde. L-Carnitin ist ein vitaminähnlicher Stoff, der mit den Vitaminen der B-Gruppe verwandt ist. L-Carnitin ist für die normalen Lebensprozesse des Organismus zur Leistungserhaltung erforderlich. Marathon gibt Ihnen Kraft und Energie, entspannt und fördert Ihr Wohlbefinden nach getaner Arbeit. Der edle und natürliche Pilsgeschmack wird Sie begeistern. Ihrer Gesundheit zuliebe verwenden Sie Marathon sorgfältig, es enthält Alkohol. Vermeiden Sie Marathon als Verkehrsteilnehmer, Schwangere, Kranker und Jugendlicher. [07.2]

Die beiden Wellness Biere *Anti Aging Bier und Marathon* der brandenburgischen *Klosterbrauerei Neuzelle* sprechen mit ihren Produktinformationen nicht offensichtlich ein bestimmtes Geschlecht an. Die Etikettengestaltungen mit Gemäldeausschnitten eines regionalen Künstlers adressieren die Biere jedoch eindeutiger.

Der erschaute Mensch ist entsprechend der Bedeutung ein Mensch, der Großes und Erhabenes wahrnimmt. Folgen wir den umfangreichen Informationen über die Inhaltsstoffe und ihrer Wirkung auf der Internetseite der Firma, so scheint der esoterisch anmutende, geschlechtslos dargestellte Mensch auf dem Etikett wohl von der Anti Aging Wirkung erschaut zu sein.

Im Gegensatz dazu, wird auf dem Etikett des *Marathon* Bieres die mit Blautönen gefüllte Silhouette eines biertrinkenden, muskulösen Männeroberkörpers gewählt. Die Produktbeschreibung enthält nur die nötigsten Informationen über die Wirkung der Inhaltstoffe und den Geschmack. So wird eine einfache und klare Aussage getroffen. ■

26 Etikett ›Anti Aging Bier‹
26 Label ›Anti Aging Bier‹
27 Etikett ›Marathon‹
27 Label ›Marathon‹

26

Marathon, the beer for afterwards. Marathon is not just an indulgence drink: it is more. Marathon is a pils beer to which L-carnitine is added. L-carnitine is an ingredient related to the B-group vitamins. L-carnitine is required to maintain the performance of the body's normal life processes. Marathon gives you power and energy, relaxes and enhances your well-being after your body has done some work. The sophisticated and natural pils taste will thrill you. For the sake of your health, use Marathon prudently, it contains alcohol. Do not drink Marathon if you intend to drive. It should also be avoided by pregnant women, people who are ill, and youngsters. 07.2

27

The two wellness beers – *Anti Aging Beer* and *Marathon* brewed by *Klosterbrauerei Neuzelle* in Brandenburg, do not obviously seem to target any specific gender in their product information. However, the design of the labels featuring extracts from paintings by a local artist, stamp the beer much more unequivocally.

The person who has seen the light is accordingly a person who has become aware of something grand and sublime. When we take another look at the comprehensive information on the ingredients and their effects as presented on the company's website, the esoteric looking genderless person on the label would appear to have seen the light of the anti-aging effect.

By way of contrast, the label for *Marathon* beer features a silhouette filled out with various shades of blue, and resembling a beer-drinking, muscular male torso. The product description only contains the minimum information necessary on the effect of the ingredients and the taste. It gets across a clear and simple message in this way. ■

08 B.United Jeans Woman; Benetton B.United Jeans Man; Benetton

◂ 36

Zwei Düfte, die sich gut tragen lassen: sinnlich und dynamisch, wie Blue Jeans. [08.1]

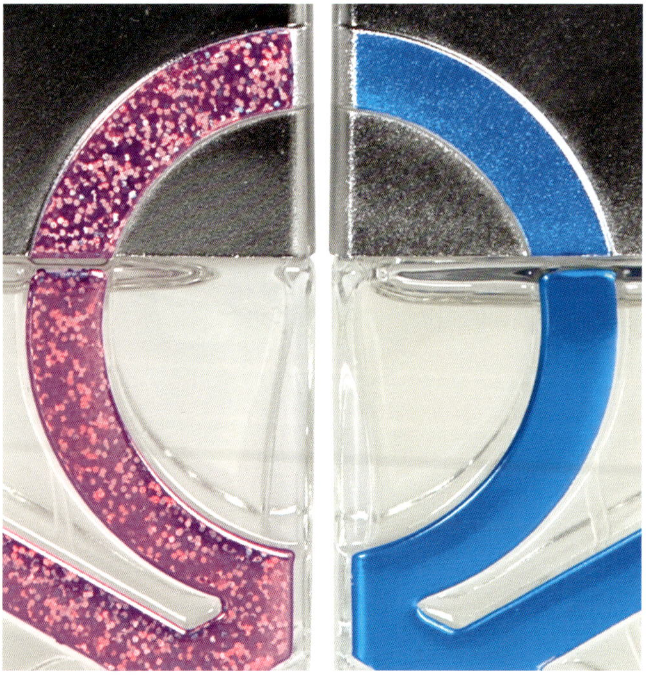

28

Die Verpackungsgestaltung des Duftes *B. United* von *Benetton* unterscheidet sich bei Damen- und Herrendüften lediglich in Details. Identische Flakons, inklusive Deckel, sind für das jeweilige Geschlecht mit einer Farbapplikation in schimmerndem Pink beziehungsweise metallischem Blau versehen. Das angeschnittene Firmensignet ist in Glas und Verschluss erhaben geprägt. Nebeneinander platziert ergeben beide Flakons das vollständige *Benetton*-Signet – Mann und Frau werden hier als zusammengehörig und gleichberechtigt dargestellt.

Mit der Verwendung eines formal geschlechtsneutral gestalteten Flakons umgeht das Unternehmen außerdem die aufwändige Produktion unterschiedlicher Verpackungen und verzichtet auf eine Wertung der Geschlechter durch die Verwendung unterschiedlicher Formen. ■

29

Two fragrances easy to wear, sensual and dynamic just like a pair of jeans. 08.1

The design of the packaging for the *B. United* fragrance from *Benetton* differs only in terms of details between the men's and women's fragrances. The identical flacons, including the top, are decorated with a colour application in glittering pink or metallic blue for the different sexes. The halved company logo is sophisticatedly engraved into the glass and the top. When placed next to one another, the two flacons reveal the complete *Benetton*-logo – man and woman are presented here as belonging to one another and enjoying equal status.

By using a formal flacon designed with a neutral (male/female) appearance, the company also saves itself the expensive production of different packaging, and avoids any valuation of the sexes by way of different shapes. ■

28 Detail Flakons ›B.United‹

28 Detail flacons ›B.United‹

29 Verpackung ›B.United‹

29 Packaging ›B.United‹

09 Pearl & Beauty, Double Effect; Nivea
Dry Impact, Cool Kick; Nivea

◂ 38

Die Verpackungen der Damen- und Herrenpflegeprodukte unterscheiden sich formal nur in Details. Lediglich das Spray für die Dame ist tailliert und weist Rundungen mit größeren Radien auf. Hinsichtlich der Namens- und Farbgebung sowie der verwendeten Motive könnten die Produkte für Frau und Mann jedoch kaum unterschiedlicher sein.

Milde Pflege, Violet Senses und Perlenextrakte sorgen bei *Pearl & Beauty* und *Double Effect* für schöne Haut. Die hochglänzenden Rosatöne der Verpackung wirken zurückhaltend, sanft und bescheiden. Dies wird durch die Verwendung von Minuskeln und die Motivwahl sowie die spielerische Anordnung der grafischen Elemente unterstützt.

Extra Frische Kick und *Cool Care Formel* beim Roll-On-Produkt *Cool Kick* und *Mineralien* beim *Dry Impact* Spray sorgen für die nötige Pflege. Oberflächen mit metallischem Glanz, simulierte matt geschliffene Stahltasten und Griffflächen. Rote und weiße Majuskeln stehen kontrastreich auf dunklen, kühlen Blau- und Grautönen. Die zentrierte Anordnung der grafischen Elemente unterstützt zusätzlich das sehr bestimmte und selbstbewusste Auftreten des Produktes. ■

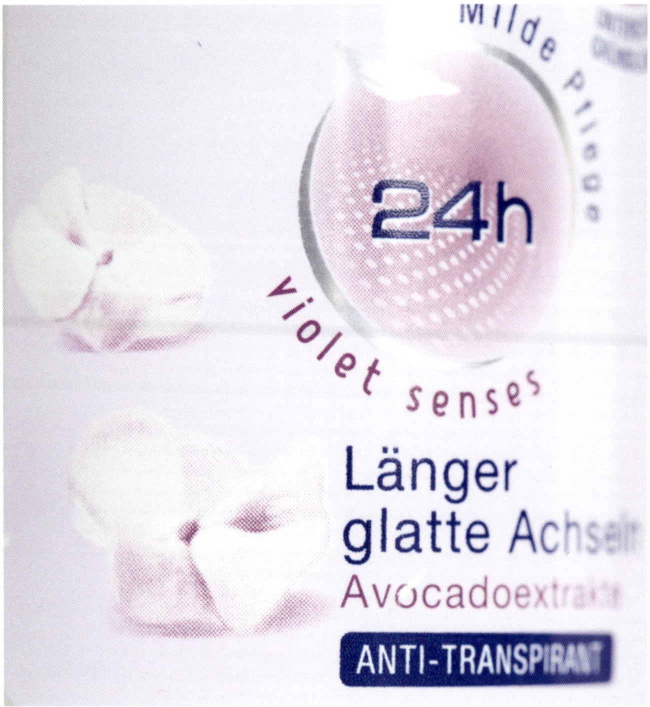

30

30 Detail ›Pearl & Beauty‹

30 Detail ›Pearl & Beauty‹

31 Verpackung ›Pearl & Beauty‹, ›Dry Impact‹

31 Packaging ›Pearl & Beauty‹, ›Dry Impact‹

32 Produktgrafik

32 Product graphic

underlined by the use of lower case, the choice of image, and the playful layout of the graphic elements.

31 *Extra Fresh Kick* and *Cool Care Formula* in the *Cool Kick* roll-on product and *Minerals* in the *Dry Impact* spray provide the care promised by the product. Surfaces with a metallic sheen, simulated matt-polished steel buttons, and gripping surfaces. Red and white capitals stand out in strong contrast to the dark and cool shades of blue and grey. And the centralised arrangement of the graphic elements stresses further the very emphatic and self-confident look exuded by the product. ■

The packaging of the women's and men's care products differ formally only in a few details: the lady's spray is waisted and has curves with large radii. In terms of their names and colour schemes, as well as the associated images, there is an enormous difference between the men's and women's products.

Gentle Care, Violet Senses and Pearl Extracts in *Pearl & Beauty* and *Double Effect* make sure the skin stays beautiful. The high-gloss pink colours of the packaging appear unassuming, soft and modest. This effect is

32

10 Vital f; Orthomol
Vital m; Orthomol

◂ 40

Orthomol Vital f ist ein diätetisches Lebensmittel für besondere medizinische Zwecke (bilanzierte Diät). Orthomol Vital f zur diätetischen Behandlung von Frauen mit chronischer Erschöpfung, Burn-out-Syndrom und stressbedingten Erkrankungen mit metabolischer Störung. [10.1]

Orthomol Vital m ist ein diätetisches Lebensmittel für besondere medizinische Zwecke (bilanzierte Diät). Orthomol Vital m zur diätetischen Behandlung von Männern mit chronischer Erschöpfung, Burn-out-Syndrom und stressbedingten Erkrankungen mit metabolischer Störung. [10.2]

33 Verpackung ›Orthomol Vital f‹
33 Packaging ›Orthomol Vital f‹
34 Verpackung ›Orthomol Vital m‹
34 Packaging ›Orthomol Vital m‹

33

Die Produkte unterscheiden sich bis auf ihre Bezeichnung und eine zurückhaltende Farbcodierung ausschließlich in der Zusammensetzung der Inhaltsstoffe, die jeweils auf die körperlichen Bedürfnisse des biologischen Geschlechts abgestimmt sind. Auf weitere gestalterische Kennzeichnungsmerkmale bei Verpackung oder Darreichungsform wird hier verzichtet. ■

Orthomol Vital f is a dietary food stuff for special medical purposes (balanced diet). Orthomol Vital f for the dietary therapy of women with chronic fatigue, burnout syndrome and stress-related illnesses involving metabolic problems. [10.1]

Orthomol Vital m is a dietary food stuff for special medical purposes (balanced diet). Orthomol Vital m for the dietary therapy of men with chronic fatigue, burnout syndrome and stress-related illnesses involving metabolic problems. [10.2]

The only difference in the products, with the exception of their names and the rather restrained colour coding, is the composition of the ingredients which are tailored to the physical needs of the biological sexes. Additional design identifying characteristics in packaging or dosage form are completely dispensed with here. ■

34

11 Venus Breeze; Gillette Mach 3; Gillette

www.gillettevenus.de: Willkommen am Strand

◄ 42

Gillette Venus – Erwecke die Göttin in Dir
Das Geheimnis einer schnellen und einfachen Rasur? Venus® Breeze™. Einfach die Rasierklinge anfeuchten und wunderbar glatte Haut genießen. Die flexiblen, integrierten Rasiergel-Kissen enthalten reichhaltige Öle, die einen feinen Schaum erzeugen, die Klinge sanft über die Haut gleiten lassen und zusätzliches Rasiergel oder Seife überflüssig machen. Integrierte Rasiergel-Kissen machen die Rasur noch einfacher. Der 3-Klingen-Rasierkopf sorgt für eine sensationell glatte Rasur. Austauschbare Ersatzklingenköpfe, Shaving Compact zur Aufbewahrung von Ersatzklingen in der Dusche, weicher Gummigriff für sicheren Halt. [11.1]

Welcher Rasierer ist der Richtige für meine Tochter? Seien Sie beruhigt. Venus® Rasierer wurden speziell für Frauen entwickelt und können auch von Mädchen sicher und einfach verwendet werden. Testen Sie unsere Venus® Rasierersuche – sie hilft Ihnen und Ihrer Tochter, den besten Rasierer für ihre Bedürfnisse auszuwählen. [...] Wie kann ich meiner Tochter helfen, dem Gruppendruck zu widerstehen, wenn es um die Rasur geht? [...] Erinnern Sie Ihre Tochter daran, dass es um ihren Körper geht und dass die Entscheidung, wann sie mit dem Rasieren beginnt, ganz bei ihr liegt. Es gibt keinen »richtigen« Zeitpunkt, um mit dem Rasieren anzufangen. Und versichern Sie ihr natürlich, dass sie sich auf Ihre Hilfe verlassen kann, wenn sie bereit ist. [11.2]

Sanft animierte Bilder, Pastelltöne und Farbverläufe erzeugen eine ›weibliche‹ Anmutung und gleichzeitig ein Gefühl von Sicherheit, Diskretion und Leichtigkeit. Im virtuellen Wellness Badezimmer: *Erfahren Sie hier, wie nah Sie Ihrem Partner wirklich sind* und *Finden Sie hier heraus, wie Sie Ihrem Partner näher kommen können*. Schönheitstipps und Stilberatung animieren die moderne Frau, ihrem Partner als betörendes und immer perfekt rasiertes Wesen gegenüberzutreten. ►

35

www.gillettevenus.de: Welcome to the beach

Gillette Venus – Reveal the goddess in you
The secret of a fast and simple shave? Venus® Breeze™. Just moisten the razorblade and enjoy wonderful smooth skin. The flexible, integrated shave gel cushion is generously endowed with oils which produce fine foam, and allow the razorblade to slide gently over the skin – making additional shaving gel or soap superfluous. The integrated shave gel cushion makes shaving even simpler. The 3-blade shaver head guarantees a sensational smooth shave. Exchangeable replacement razor heads, Shaving Compact to store replacement blades in the shower, soft rubber handle for a secure grip. [11.1]

36

Which razor is the right one for my daughter? Relax. Venus® razors were specially developed for women and can also be safely and simply used by girls. Test our Venus® razor search – it will help you and your daughter to choose the best razor for your requirements. […] How can I help my daughter resist peer pressure with respect to shaving? […] Remind your daughter that it is her body, and that it is her decision alone to determine when she wants to start shaving. There is no »proper« time when shaving should begin. And reassure her that she can naturally rely on your support when she decides that the time has come. [11.2]

Softly animated pictures, pastel shades and colour fading create a ›feminine‹ look, at the same time as a feeling of security, discretion and lightness. In the virtual spa bathroom: *find out here how close you really are to your partner and find out how you can get closer to your partner.* Beauty tips and style advice motivate today's modern woman to be seen by her partner as a beguiling and always perfectly shaved being. ▸

35 Verpackung ›Gilette Venus Breeze‹
35 Packaging ›Gilette Venus Breeze‹
36 Verpackung ›Gilette Mach 3‹
36 Packaging ›Gilette Mach 3‹

Jedes Jahr schneller, schärfer und präziser rasieren! So jagt eine Produktserie die nächste:
Der Mach 3 Turbo - Turboleistung.
Der Gillette Mach 3 Turbo vereint ausgeklügelte Rasiertechnologien, die als nicht batteriebetriebenes System für eine gründliche und komfortable Rasur sorgen. Hol dir die Turboleistung für deinen Style und Höchstleistung für deine Rasur. Der Gillette Mach 3 Turbo vereint die 3-Klingen-Rasiertechnologie mit den Anti-Friction-Klingen, die sanft über dein Gesicht gleiten. Sein beweglicher Klingenkopf mit 10 hauchdünnen Lamellen folgt jeder Kontur deines Gesichts und sorgt so für unsagbaren Komfort und Gründlichkeit bei der Rasur. [...] Der Lubrastrip des Rasierers verblasst, sobald die optimale Rasierleistung nicht mehr gegeben ist. [11.3]

37 Werbedarstellung ›SB Beauty‹
37 Commercial ›SB Beauty‹
38 Werbedarstellung ›Nassrasur mit Fusion ProGlide Power 3‹
38 Commercial ›Wet shave with Fusion ProGlide Power 3‹

Auf der Website wird es wissenschaftlich: Ein umfangreiches Rasur-Lexikon in der Rubrik Men´s Style führt von Aftershave bis Zwirbelbart durch die Welt des Rasierens – selbstverständlich mit zahlreichen Hinweisen auf Gillette Produkte. Kräftiges Blau, kontrastreiches Gelborange verbunden mit Elementen des CAD – wie Männerköpfe als Drahtgittermodelle – führen zu einer technisch wirkenden Gestaltung. Dynamische Animationen vermitteln das Bild des technikaffinen und sportlich modernen Mannes, der diesen Nassrasierer benötigt. [11.4] ■

37

38

Faster, sharper and more precise shaving from year to year! Each new product series is launched hot on the heels of the last:
The Mach 3 Turbo – Turbo power. The Gillette Mach 3 Turbo marries sophisticated razor technologies to ensure a thorough and comfortable shave even without batteries. Get the turbo power for your style and high power for your shave. The Gillette Mach 3 Turbo combines 3-blade razor technology with anti-friction blades which slide smoothly over the skin. Its supple razor head with 10 wafer-thin lamellae follows every contour of your face to give you an inexpressibly comfortable and thorough shave. [...] The Lubrastrip in the razor becomes paler as soon as the optimal razor performance declines. [11.3]

The website gets scientific: a comprehensive grooming glossary in the *Men´s Style* section covers every aspect of shaving from aftershave to wet shaving – obviously with plenty of references to *Gillette* products. Muscular blue, strongly contrasting orange-yellow, combined with CAD elements – such as male heads as grid models – create a strong technical feel. Dynamic animations promote the image of a modern hitech-savvy and sporting man who needs this wet razor. [11.4] ■

39

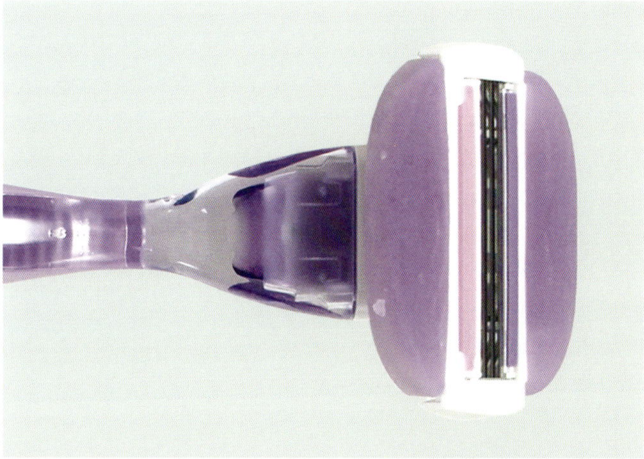

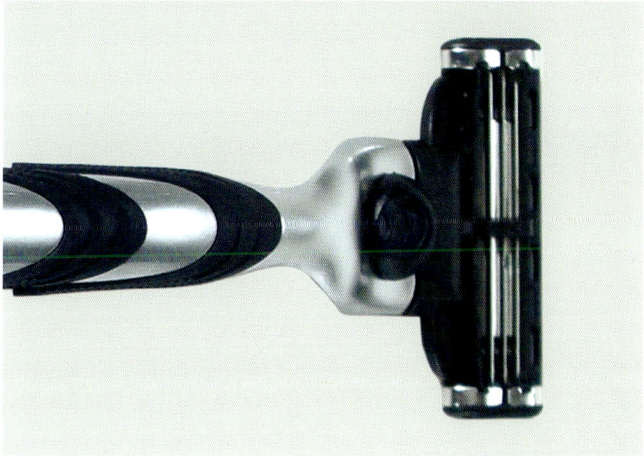

40

39 Rasierkopf ›Gillette Venus Breeze‹

39 Razor head ›Gillette Venus Breeze‹

40 Rasierkopf ›Gillette Mach 3‹

40 Razor head ›Gillette Mach 3‹

12 Showergelserie; Dove
Showergelserie; Dove Men+Care

◀ 44

Clean Comfort, Extra Fresh und *Aqua Impact* heißen die Pflegeprodukte der *Dove Men+Care* Serie. Bereits die Namensgebung charakterisiert die Produkte als aktive und effiziente Helfer. Die formale Geradlinigkeit durch die auf dem Verschluss stehende Duschgelverpackung wirkt nicht statisch, da die Form durch horizontal verlaufende schwungvolle Fugen unterbrochen wird. Die Majuskeln sowie die Produktgrafik unterstützen den Formcharakter. Kontrastreich stehen die grafischen Elemente in frischen Grün- und Blautönen auf der dunkelgrauen, metallisch schimmernden Verpackung. Silberne Farbakzente führen zu einer technischen Anmutung des Produktes.

Dove MEN+CARE wurde für die speziellen Bedürfnisse männlicher Haut entwickelt. Die hochwirksame Formel mit MICROMOISTURE™ liefert intensive Pflege und Frische, die Haut wird mild gereinigt und Hauttrockenheit nach dem Duschen effektiv bekämpft. [12.1] ▶

41 ›Dove‹ Duschgelserie
41 ›Dove‹ shower gel series
42 ›Dove Men+Care‹ Duschgelserie
42 ›Dove Men+Care‹ shower gel series

42

Clean Comfort, Extra Fresh and *Aqua Impact* are the names of the care products in *Dove's Men+Care* series. The names themselves label the products as active and efficient helpers. The formal straightness of the shower gel packaging which stands upside-down on its top, does not have a static appearance because the shape is broken up by curving horizontal grooves. The capital lettering and the product graphics emphasise the form character. The graphic elements in fresh shades of green and blue contrast with the dark-grey, metallic shimmering packaging. Silver coloured highlights give the product a high-tech look.

Dove MEN+CARE was developed for the special needs of male skin. The highly effective formula with MICRO-MOISTURE™ provides intensive care and freshness, the skin is gently cleaned, dry skin after showering is effectively prevented. [12.1] ▸

Bekämpft wird bei den Damenprodukten von *Dove* nichts. Hier gibt es laut Etikett vielmehr *Belebende Mikroperlen, die die Haut wachküssen.* Oder es wird mit einer *reichhaltigen feuchtigkeitsspendenden Creme-Dusche, weiche Haut gezaubert, luxuriös verwöhnend und eine blumige Verführung.*

Die Form der Duschgelverpackung ist fließend, weich und tropfenförmig. Die Verpackung hat einen skulpturalen und emotionalen Charakter. Der Dosierkopf mit Verschluss ist auffallend klein und im Vergleich zur Basis stark verjüngt. Farbwahl und Motive der Abbildungen sind zart und zurückhaltend und stehen auf mattweißem Untergrund. Anders
43 als bei der *Men+Care* Serie wird auf den Einsatz von Majuskeln verzichtet und eine zurückgenommene und bescheiden wirkende Kleinschreibung verwendet. In der Gesamterscheinung wirkt die Frauenserie von *Dove* anmutig und nahezu schwerelos. ■

43 Detail ›Dove Creme-Öl Beauty Pflegedusche‹

43 Detail ›Dove Cream-Oil Beauty Nourishing Body Wash‹

44 Produktgrafik ›Dove Men+Care‹ Duschgelserie

44 Product graphic ›Dove Men+Care‹ shower gel series

Dove's women's products do not fight anything. Instead, the label here promises *revitalising micro-pearls which kiss your skin awake. Or a rich moisturising cream shower gel, conjures up a soft skin, pampers luxurously, and promises a flowery seduction.*

The shower gel packaging has a flowing shape, soft and drop-like. The packaging has a sculptural and emotional character. The dosing head with the cap is remarkably small and very much narrower than the base. The choice of colours and images in the visuals is delicate and discreet, and stand against a white matt background. Unlike the *Men+Care* series, instead of using capital letters, use is made of discreet and modest-looking lower case lettering. The overall appearance of *Dove's* women's series is attractive and almost weightless. ■

44

13 Contradiction; Calvin Klein
Contradiction for men; Calvin Klein

◂ 46

45

46

Unerwartet, extravagant und provokant – Contradiction zelebriert die Erotik der Widersprüchlichkeit, die jeder Frau innewohnt. Strukturierte Klarheit verbindet sich mit kopfloser Sinnlichkeit zum perfekten Ausdruck wahrer Feminität. [13.1]

Maskulin, stark und doch sinnlich und gefühlvoll – Contradicion Men ist ein Duft voller Kontraste, der kühle Spritzigkeit mit Kräuternoten und exotisch-holzigen Aromen verbindet. Aufregender wurde die Anziehungskraft des Widersprüchlichen selten in Szene gesetzt. [13.2]

Beide Flakons basieren auf geometrischen Grundkörpern. Der Duft für die Dame – klassisch rund als geschlossener Zylinder, der Duft für den Mann – eckig als Quader. Die überproportionalen Deckel verbunden mit dem Materialkontrast zum Glasbehälter verleihen den Flakons trotz der Einfachheit eine große Eigenständigkeit. Der in einem warmen Hautton gefärbte Damenduft tritt durch den dünnwandigen Glasflakon offen, fast duftend hervor. Der gewölbte Innenboden und der seidenmatt gehaltene strukturierte Deckel unterstützen den weichen Eindruck. An Eiswürfel erinnernd der Männerduft: robust und kühl. Der matte Deckel, die starke Glaswandung und die fast wässrige Farbe des Duftes unterstützen den Charakter des Produktes.

Die Karton-Umverpackungen beider Produkte unterscheiden sich lediglich in der Farbgebung und der Folienschriftprägung des Herrenduftes: Silber wird hier dem Damenduft und Anthrazit dem Herrenduft zugeordnet. ■

47

Unexpected, extravagant and provocative – Contradiction celebrates the eroticism of contrariness which dwells in every woman. Structured clarity is joined by unthinking sensuality to perfectly express true femininity. [13.1]

Masculine, strong but also sensuous and emotional – Contradiction Men is a cologne full of contrasts which combines cool liveliness with herbal hints and exotic woody aromas. The attraction of contradictions has rarely been captured in such an exciting way before. [13.2]

Both flacons are based on basic geometric shapes. The women's perfume – classically round as a closed cylinder; the cologne for men – angular as a brick. The over proportional size of the tops, combined with the contrasting materials compared to the glass container lend the flacons a great deal of autonomy despite their simplicity. The women's fragrance in warm skin colour can be clearly seen – almost fragrantly – through the thin walls of the glass flacon. The arched internal base and the silk-matt structured lid emphasise the soft impression. The men's cologne has anaogies with a cube of ice: robust and cool. The matt top, the thick glass walls, and the almost watery colour of the fragrance emphasise this character of the product.

The cardboard packaging around both products only differs in terms of colour and the embossed film of the men's fragrance: silver is used here for the women's perfume and anthracite for the men's cologne. ■

45 Materialität Flakons gechlossen
45 Materiality closed flacons
46 Materialität Flakons geöffnet
46 Materiality closed flacons
47 Verpackung
47 Packaging

14 Telephone Plush pink; Silly Gifts, Present Time Telefon T3, Jacob Jensen

◂ 48

48

Diese so ungleichen Produkte wenden sich natürlich an unterschiedliche Käufergruppen und sind in verschiedenen Preissegmenten angesiedelt. Die Produktgegenüberstellung der Telefone macht jedoch die genderspezifische Unterscheidung durch Farbe, Material und Verarbeitungsqualität exemplarisch deutlich.

Die weiblich konnotierten Gestaltungselemente der Farbe Pink und des Materials Plüsch ergeben in Verbindung mit dem glänzenden Kunststoff einen minderwertigen, eher unbrauchbaren Eindruck.

Das Pendant bedient sich in der Gestaltung eher einer funktionalen Formsprache. Materialauswahl, Verarbeitung und Gebrauch vermitteln Hochwertigkeit und das Ergebnis scheint durchdachter und professioneller. Die Schwere, die durch Farbgebung und Form erreicht wird, wirkt seriös und insgesamt maskulin.

Voll funktionsfähiges Telefon für analogen Anschluss. Das Telefon lässt sich an jede Anlage anschließen, egal ob Sie über eine ISDN, DSL oder analoge Zuleitung verfügen.
Kuschelig weich verpackt in rosafarbenen Plüsch kommt dieses Telefon daher. Neben seinen flauschigen Applikationen besticht es durch sein sympathisches Old-School-Design. Damit wird jedes Telefonat zu einem ganz besonderen Gespräch. Einfach in die analoge Anschlussdose damit, und schon geht's los! Hat's geklingelt? [14.1]

Die Telefone – designed von Jacob Jensen – sind einzigartige funktionale Produkte. Der Hörer hat eine elegante schlanke Form und hebt sich durch sanften Druck auf den oberen Teil des Hörers direkt in die Hand des Benutzers. Die Telefone sind in wenigen Worten das typische Ergebnis der Denkweise Jacob Jensens: Unaufdringlich und elegant in einer perfekten Balance zwischen Form und Funktion. Mit dem klassischen Jacob Jensen Design werden Telefonverbindungen zu einem besonderen Vergnügen. [14.2] ■

48 Detail ›Plüschtelefon‹
48 Detail ›plush telephone‹
49 Detail ›Telefon T3‹
49 Detail ›Telephone T3‹

Products with such obvious differences are naturally aimed at different purchaser groups and are located in different price segments. However, comparing the telephones highlights the genderspecific differentiation using colour, material and the quality of the workmanship.

The female connotations of design elements involving the colour pink and the plush material, combined with the glossy plastic, create a cheap, maybe even unusable impression.

The contrasting object is based on a design incorporating rather more functional elements. The choice of material, workmanship and function get across a sense of high quality, and the result appears to be better thought out and more professional. The solidity created by the colour scheme and the shape produce a serious and overall masculine effect.

49

Fully functioning telephone for analogue connection. The telephone can be connected up to every system, and is not fussy whether you have an ISDN, DSL or analogue connection. This telephone is packaged cuddly and soft in a pink-coloured plush material. In addition to its fluffy decorations, it also boasts a sympathetic old school design. All of this turns every telephone call into a very special conversation. Just connect it up to your analogue socket and away you go! Does that ring a bell with you? [14.1]

The telephones designed by Jacob Jensen are unique functional products. The handset has an elegantly slim shape and lifts up directly into the hand of the user by gently pressing on the upper part of the handset. The telephones are the typical result of Jacob Jensen's philosophy, and can be summed up in just a few words: unostentatious and elegant with a perfect balance between form and function. The classic Jacob Jensen design turns a telephone into a special pleasure. [14.2] ■

15 LEGO® Bauspaß für Mädchen LEGO® Technic

◂ 50

50

Lass dich von den tollen Farben in diesem Set inspirieren und baue ein kleines Pony oder ein tolles Haus für dich und deine Freunde! Lass deiner Fantasie mit dieser LEGO Steinbox freien Lauf. Die Spezialelemente in Mädchenfarben lassen keine Wünsche offen. [15.1]

Das Go-Kart mit vielen Details bringt Spaß auf der Rennstrecke. Gib Gas und lege dich in die nächste Kurve. Mit vielen bekannten Technic Elementen, wie beweglichem Kolben und Lenkung. Das Modell lässt sich in einen Aufsitz-Rasenmäher umbauen. [15.2]

LEGO setzt neben den für alle Kinder konzipierten Produkten mit der Sparte *LEGO Technic* auf die Adressaten der männlichen Kunden. Sowohl in der Modellauswahl (Baufahrzeuge, Motorräder, Trucks...) als auch in deren Gestaltung und Beschreibung wird der technikinteressierte Junge im Alter von 8 bis 14 Jahren angesprochen.

Für Mädchen gibt es in diesem Sinne keine entsprechend attraktiven Modellangebote in der *Technic* Sparte. Hier wird davon ausgegangen, dass sie sich für eine heimelige Welt rund um Haus und Hof, Schlösser und Feen oder Tiere interessieren (Produktlinie *Belleville*). Die Farbgebung von Produkt und Verpackung ist stereotyp: Rosa, Pink und Lila. Die Figuren erinnern in ihren Charakteren an die Barbiewelt. Sie stehen damit im Kontrast zu den sonst für *LEGO* typischen, geometrisch aufgebauten Figuren. Die Gestaltungsmöglichkeiten der Mädchenserie sind wesentlich geringer als beim Spielen mit den Angeboten aus anderen Produktserien, wie beispielsweise *LEGO Technic*. ■

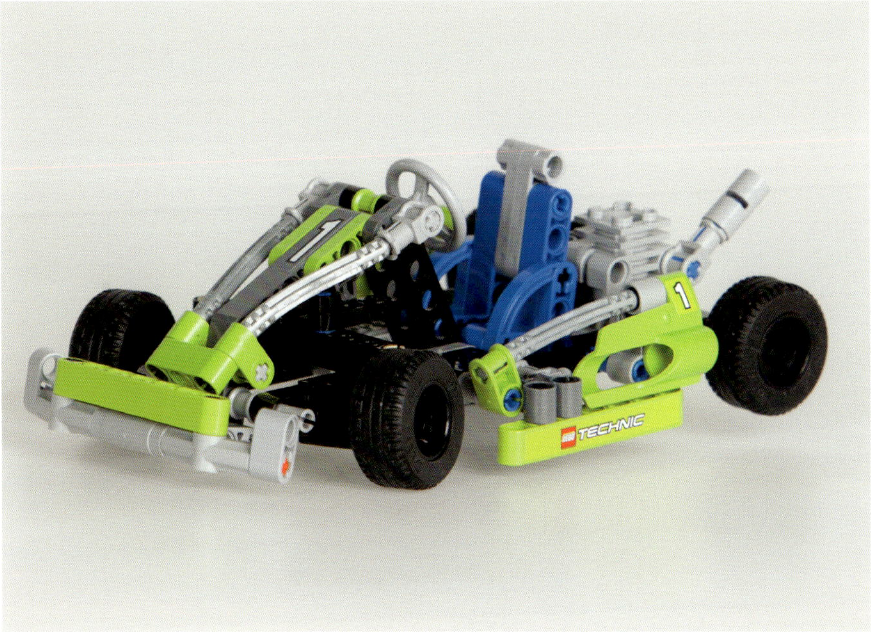

51

Be inspired by the fantastic colours in this set, and build a small pony or a lovely house for you and your friends! Set your fantasy free with this LEGO brick box in beautiful colours. The special pieces in colours girls love are just perfect. [15.1]

The go-kart with lots of details is real fun on the race track. Put your foot down and glide into the next corner. With many well-known Technic features, such as moving pistons and steering. The model can be converted into a sit-on lawn mower. [15.2]

50 Inhalt ›LEGO® Bauspaß für Mädchen‹
50 Contents ›LEGO® Pink Brick Box for girls‹
51 Inhalt ›LEGO® Technic Go-Kart‹
51 Contents ›LEGO® Technic Go-Kart‹

In addition to the products designed for all children, *LEGO* focuses with its *LEGO Technic* segment on male customers. The selection of the models (construction vehicles, motorbikes, trucks...) as well as the design and descriptions, are aimed at boys aged from 8 to 14 years interested in technology.

In this sense, there is no appropriate attractive range of models in the *Technic* segment for girls. The assumption here is that they are more girls being interested in a cosy world involving home and farm, castles and fairies, and animals (Belleville product line). The colour schemes used in the product and the packaging are stereotyped: pink and lilac. The figures resemble characters from the Barbie world. They therefore differ from the geometrically shaped figures which are otherwise typical for *LEGO*. The creative possibilities opened up by the girls series are much more restricted than the play opportunities opened up by the other product series such as *LEGO Technic*. ■

16 Hybrid; OGEL LEGO® Technic

◄ 52

52

53

Um die *Technic* Serie aus ihrer stereotypen Sparte zu befreien, wurde im Rahmen des Forschungsprojektes der Hochschule Hannover *Genderspezifische Gestaltung von Produkten und Nutzungskonzepten* ein neues Modell entwickelt: der Kinder- oder Puppenwagen als *OGEL Technic* Bausatz. Hiermit wird eine Brücke zwischen den bisher bei *LEGO* stark getrennten weiblichen und männlichen Produktbereichen gebaut. Mit diesem Bausatz wird auch Mädchen der Zugang zu einem vielfältigen und konstruktiv orientierten Spielen ermöglicht. Gleichzeitig wird das Interesse der bisher auf ›coole‹ Fahrzeuge konditionierten Jungen erweitert. ■

52 ›OGEL-Hybrid Puppenwagen‹ Verpackung
52 ›OGEL-Hybrid Doll´s pram‹ packaging
53 ›OGEL-Hybrid Puppenwagen‹ Bauvariante Buggy
53 ›OGEL-Hybrid Doll´s pram‹ buggy modell
54 ›LEGO® Technic‹ Verpackung
54 ›LEGO® Technic‹ packaging

As part of the *Gender-specific design of products and use concepts* research project conducted by the Applied Science University Hannover, a new model was developed to free the Technic series from its stereotyped segment: pushchairs and doll's prams as *OGEL Technic* modules. This bridges the strong separation between the male and female product sectors practised by *LEGO*. This buliding set also gives girls access to varied and building-oriented games. It also expands the interest of boys prevously conditioned to ›cool‹ vehicles. ■

54

◂ 54

17 Mutter; Depesche Vertrieb Papa; Depesche Vertrieb

Die collagenartige Gestaltung der Geburtstagskarten zeichnet klischeebesetzte Rollenbilder von Müttern und Vätern.

Die Karte für die Mutter erinnert eindeutig an ein Küchenszenario. Der karierte Hintergrund der Karte ähnelt dem Muster eines Tisch- oder Küchentuchs und die geblümte Küchenschürze mit einer strassbesetzten Satinschleife rundet das Bild ab. Die verwendete Schrift lehnt sich einer handgeschriebenen Schrift an. Das mit Strasssteinchen befestigte Schild mit dem Schriftzug ›Mutter‹ verweist durch seine Typografie und die angedeuteten Nähte auf den idyllisch häuslichen Eindruck der 50er Jahre.

55

Bei der Karte für den Vater ist der Text sachlich in einer breit laufenden Antiqua in Kapitälchen gesetzt und erinnert an Zeitungsüberschriften. Auch für den Schriftzug des Wortes ›Papa‹ wurde Zeitungspapier als Vorlage verwendet. Kleine metallisch anmutende Nieten heften das locker an den Kanten ausgefranste Schriftfeld auf dem schwarzweiß gestreiften Hintergrund, der an Schiedsrichtertrikots oder Pyjamastoff erinnert. ■

55 Detail Geburtstagsgrußkarte ›Mutter‹
55 Detail Birthday card ›Mother‹
56 Detail Geburtstagsgrußkarte ›Papa‹
56 Detail Birthday card ›Father‹

56

The collagen-like designs of the birthday cards show clichéd role images of mothers and fathers.

The mother's card is clearly reminiscent of a kitchen scene. The checkered background of the card resembles the pattern of a tablecloth or tea towel, and this picture is rounded off by the flowery kitchen apron with by a bling-decorated satin bow. The typeface resembles handwriting. The label with the words ›Mother‹, decorated with bling, recalls by way of its font and the suggestion of seams, the idyllic domestic images of the 1950s.

The fathers's card on the other hand has the words written soberly in capital letters in a broad Antiqua font, reminiscent of newspaper headlines. A newspaper was also used as a model for the lettering of the word ›Father‹. Small metallic-looking rivets hold the label with the slightly frayed edges onto the black-and-white striped background that resembles a football referee's shirt or pyjama material. ■

18 Bosch private collection; Theo Klein
Bosch mini; Theo Klein

Perfekt faltenfrei. Knittrige Hemden,
◄ 56 *zerknautschte Puppen-Blusen? Nicht mit dem Bosch Bügeleisen für Kinder. Das schicke, detailgetreue Gerät sieht nicht nur aus wie das Original von Mama und Papa: Die authentische Wasser-Sprühfunktion macht das Spielzeug besonders realistisch.* [18.1]

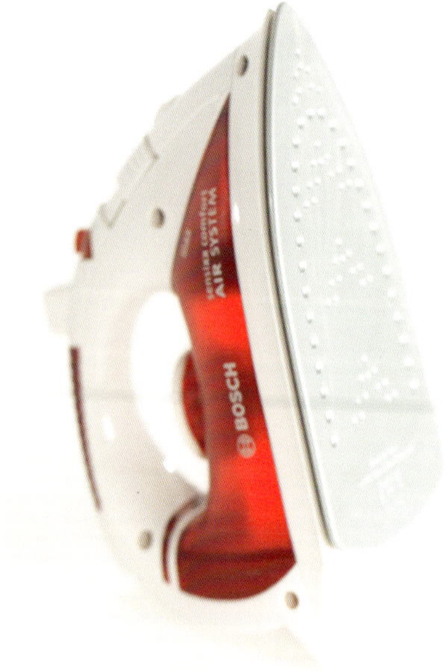

57

58

Wenn der Papa die Bohrmaschine auspackt und das neue Regal für die Diele in Angriff nimmt, möchte Klein Peter am liebsten mithelfen. Da Heimwerken aber für Kinder nicht ganz ungefährlich ist, durften sie bisher meist nur zuschauen. Damit ist jetzt Schluss, denn mit den Spielwerkzeugen Bosch mini bekommt der Nachwuchs völlig gefahrlos Kontakt mit dem Hobby Heimwerken. [...] Für alle Bosch minis gilt: Wie beim richtigen Bohren, Schleifen oder Sägen geben die Geräte die entsprechenden Geräusche von sich. In ihren grünen Kunststoffgehäusen mit rotem Bosch-Logo sind die Spielwerkzeuge ihren Vorbildern zudem zum Verwechseln ähnlich. In Sachen Sicherheit geht Bosch bei seinem pädagogischen Spielzeug keine Kompromisse ein. [...] [18.2] ►

57 Bügeleisen ›Bosch private collection‹
57 Iron ›Bosch private collection‹
58 Detail Bügeleisen ›Bosch private collection‹
58 Detail Iron ›Bosch private collection‹
59 Werbedarstellung ›Bosch mini‹
59 Advertisement ›Bosch mini‹
60 Detail Kettensäge ›Bosch mini‹
60 Detail chainsaw ›Bosch mini‹

59

When daddy unpacks the drill to start fixing new shelves in place in the hall, little Peter would just love to help. But because DIY can be dangerous for children, they have so far only been allowed to watch from a distance. No more, with the new Bosch mini toy tools, youngsters can play at doing DIY jobs in a completely safe way [...] Bosch minis have one thing in common: just like real drills, sanders or saws, the toy appliances all make the right noises. And in their green plastic boxes with red Bosch logos, the toy tools can almost be mistaken for the real thing. Bosch has been completely uncompromising when it comes to the safety of its pedagogical toys. [...] [18.2] ▸

60

Perfectly flat. Wrinkled shirts, crumpled up doll's blouses? Not with the Bosch iron for children. This smart appliance, accurate down to the last detail, not only looks just like mummy's and daddy's original: the authentic water spray function makes the toy even more realistic. [18.1]

Die Marke *KLEIN* bietet mit den Serien *Bosch private collection* und *Bosch mini* unterschiedliche Spielzeuge an, die Originalprodukten aus den Bereichen Werkzeug und Haushalt nachempfunden sind. In ihren Funktionen beschränkt, ermöglichen die Miniaturen Kindern das spielerische Erlernen typischer Arbeiten im Haushalt.

Transparente Partien machen die Funktionsweisen der einzelnen Elemente der *Bosch mini* Kettensäge sichtbar. Die Kette dreht sich in ihrem Lauf und der Schalter lässt sich wie beim Original betätigen. Hemmschwellen gegenüber technischem Gerät können so abgebaut oder bereits vor ihre Entstehung verhindert werden.

Am Bügeleisen der Serie *Bosch private collection* ist die Sprühfunktion im vorderen Bereich des Griffs funktionsfähig. Die in diesem Produkt eingesetzte Transparenz dient hier nicht der Offenlegung von Funktionsprinzipien, sondern zeigt lediglich den Wasserstand im Bügeleisen. Ein umfassender Lerneffekt über die gesamte Funktionsweise des Gerätes bleibt aus. ■

With the *Bosch private collection* and *Bosch mini* series, the *KLEIN* brand offers a range of toys which replicate the original DIY and household products. With their restricted functions, the miniature replicas enable children to playfully learn about typical jobs undertaken around the house.

Transparent parts reveal how some elements of the *Bosch mini* chainsaw function. The chain rotates and the switch can be operated just like the original. This can either breakdown the barriers with respect to technical equipment, or prevent barriers being erected in the first place.

The spray function at the front of the grip works properly in the iron in the *Bosch private collection* series. The transparent parts incorporated in this product are not there to reveal how it functions, but just to indicate the level of water in the iron. There is no broad learning effect about the overall way the appliance functions. ■

19 Pliers Combo (Pink); Tomboy Tools®
Zangen; Wiha

◂ 58

61

61 Zange ›TOMBOY TOOLS‹
61 Pliers ›TOMBOY TOOLS‹
62 Zange ›Wiha‹
62 Pliers ›Wiha‹

Die Marke *TOMBOY TOOLS®* hat zum Ziel, die Arbeit im Heimwerkersegment für Frauen zu optimieren. Es wurden spezielle Werkzeuge für die ergonomischen und physiognomischen Bedürfnisse des weiblichen Geschlechts entwickelt. Hämmer mit geringerem Gewicht und Griffe mit geringerem Volumen versprechen hier eine bessere Nutzbarkeit. Die Farbe Pink dient als ironisches Zitat bekannter Klischees und Marketinginstrumente: Färbe etwas Pink und es wird für die Frau interessant. Die Ausführung der Werkzeuge ist nicht hochwertig, was die Nutzbarkeit nachteilig beeinflusst. ▸

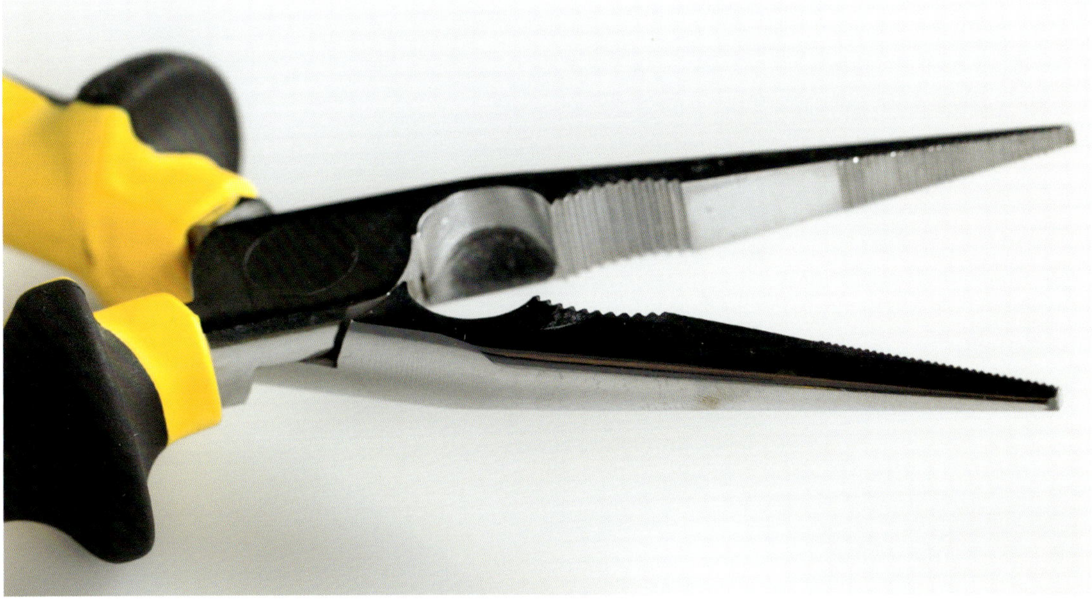

62

The *TOMBOY TOOLS®* brand aims to optimise work for women in the DIY segment. Special tools were developed to satisfy the ergonomic and physiognomic needs of the female sex. Lightweight hammers and smaller volume grips promise improved operability here. The colour pink serves as an ironic echo of well-known clichés and marketing instruments: give something a pink colour and women will be interested. The low quality construction of the tool has a negative effect on its usability. ▸

TOMBOY TOOLS® ist ein Anbieter von Werkzeugen und Heimwerkerbedarf von Frauen für Frauen – mit einer einzigartigen Produktpalette für alle Ihre Verschönerungsprojekte in Wohnung und Heim. Unser Kerngeschäft besteht darin, Frauen qualitativ hochwertige Werkzeuge, Verfahren und Techniken nahe zu bringen, und Frauen zu informierten und mutigen, heimwerkenden Wohnungs- und Hausbesitzerinnen zu machen. Das erreichen wir durch unsere begehrten In-Home Workshops – den Tool Parties!

Am Anfang waren wir lediglich eine Gruppe von Frauen, tätig im Rahmen unterschiedlicher Heimwerkerprojekte an Haus und Wohnung. Wir trafen uns, und stellten gemeinsam fest, dass wir mit den im Einzelhandel verfügbaren Werkzeugen, bzw. den üblichen Baumarktkonzepten nicht zurechtkamen. Es war und ist grundlegend zu bemängeln, dass die meisten Bau- und Heimwerkermärkte bei der Berücksichtigung des Frauenanteils unter den Kunden ein völlig antiquiertes Konzept verfolgen. Dort glaubt man, dass man, um das ideale Frauenwerkzeug zu schaffen, einfach ein Männerwerkzeug hernimmt und es in pink anbietet. Hallo?

Natürlich bieten wir in unserem Unternehmen auch Produkte in pink an – das ist jedoch auf unsere Beteiligung an der Unterstützung einer Wohltätigkeitsorganisation zurückzuführen, die sich für Frauen mit Brustkrebs engagiert; mit unseren Tools in pink wollen wir bestimmt keine Klischees bedienen.
Unser Unternehmen ist eben anders als alle Ihnen bekannten Baumärkte und Einzelhändler für den Heimwerkerbedarf. Wir fördern und unterstützen die Aus- und Weiterbildung, und unsere Unternehmenskultur unterstützt Frauen dabei, zu selbstbewussten, kompetenten und unabhängigen Wohnungs- und Hausbesitzerinnen zu reifen – das ist unsere ganz neue Definition von TOMBOY! [19.1]

Die Werkzeuge von *Wiha* erfüllen die ergonomischen Anforderungen und strahlen in ihrer Farbgebung sowie der Verarbeitung von Griffpartien und Zangenköpfen Professionalität aus. ■

TOMBOY TOOLS® is a tool and home improvement company for women, by women, providing a unique range of tools for home improvement projects. Our primary business is to provide the highest quality tools and techniques to women, empowering them to become confident and competent homeowners through our In-Home Workshops – Tool Parties!

In the beginning, we were just a group of women working on various home improvement projects that got together to lament the fact that retail hand tools and those warehouse workshops don't work for us. The simple fact is that most hardware companies are antiquated in their approach to women. Most figure that in order to appeal to women, you just take a tool made for a man and make it pink. Hello?

We do have a pink range but this was born out of our desire to support a breast cancer charity, rather than reinforcing stereotypes. Our company is very different from the run of the mill home improvement retailers. We advocate education and have an internal culture that supports women in becoming confident, competent homeowners – our new definition of TOMBOY! [19.1]

The *Wiha* tools satisfy the ergonomic demands, and their colouration and the finishing of the grips and the jaws exude professionalism. ■

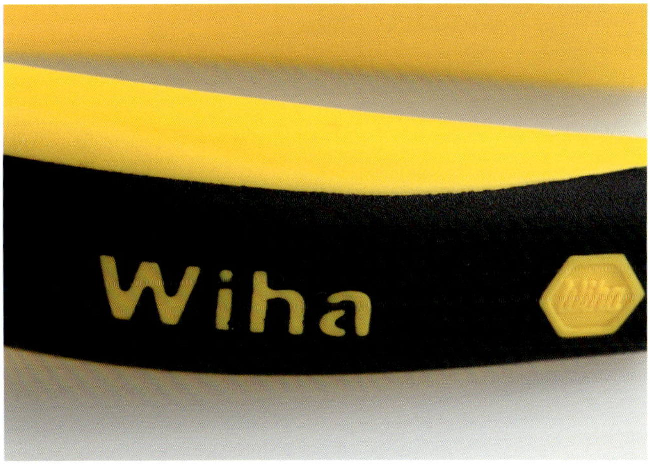

64

63 Detail Zange ›TOMBOY TOOLS‹
63 Detail Pliers ›TOMBOY TOOLS‹
64 Detail Zange ›Wiha‹
64 Detail Pliers ›Wiha‹

20 Vileda junior; Theo Klein Bosch mini; Theo Klein

◂ 60

Das größte Gut, das Eltern ihren Kindern nach der Elternliebe und häuslicher Geborgenheit mitgeben können, ist eine Umgebung, die soziale, praktische, gefühlsmäßige und geistige Entwicklung ihrer Kinder fördert. Dazu gehört auch gutes Spielzeug. Gerade im Spiel können Kinder ihre Neugier, Phantasie, Kreativität und ihr praktisches Wissen gut entwickeln.

Dabei ist es wichtig, dass das Angebot an Spielzeug so ausgeglichen ist, dass alle, für die Entwicklung des Kindes wichtigen Säulen seiner Persönlichkeit gefördert und weiterentwickelt werden. Die Kollektion von KLEIN wird diesem Anspruch gerecht. Sie umfasst Spielzeuge, die für die Entwicklung Ihres Kindes besonders wertvoll sind. […] [20.1]

Die Marke *KLEIN* bietet mit den Serien *Bosch mini* und *Vileda junior* unterschiedliche Spielzeuge an, die Originalprodukten nachempfunden sind. In ihren Funktionen beschränkt, ermöglichen die Miniaturen Kindern das spielerische Erlernen typischer Arbeiten im Haushalt, deren Ausführung sie im Alltag bei den Eltern beobachten. ▸

65 Detail ›Vileda junior‹
65 Detail ›Vileda junior‹
66 Detail ›Vileda junior‹
66 Detail ›Vileda junior‹

65

66

The greatest thing that parents can give their children, apart from parental love and domestic security, is an environment which supports the social, practical, emotional and spiritual development of their children. This also includes good toys. Play in particular is the ideal way for children to properly develop their curiosity, fantasy, creativity and practical knowledge.

It is important in this context that a balanced range of toys is available to ensure that all of the important aspects of a child's personality vital for its development are properly nurtured and further developed. The KLEIN collection satisfies these criteria. It includes toys that are particularly valuable for the development of your child. [...] [20.1]

With the *Vileda junior* and *Bosch mini* series, the *KLEIN* brand offers a range of toys which replicate the original products. With their restricted functions, the miniature replicas enable children to playfully learn about typical jobs undertaken around the house by their parents. ▸

67

Mit einer speziellen Edition für Kinder bringt Vileda frischen Wind in den Spielalltag. […] Zukunftsorientierte Entwicklungen und starke Qualitätsprodukte sind stellvertretend für den Markennamen Vileda. Gewachsenes Erfahrungspotenzial geht Hand in Hand mit einer zukunftsorientierten Produkt- und Markenstrategie. Und hier liegt unser strategischer Ansatz: Kinder sind unsere Kunden von Morgen. Das positive Markenimage, das wir schon früh in Form von kindgerechtem, hochwertigem Spielzeug prägen, ist ein starker Beitrag zur Emotionalisierung der Marke Vileda und unserer Produkte. […] Mit Liebe zum Detail sind alle wesentlichen Produkteigenschaften der Vileda Miniaturausgaben ihrem Originalvorbild nachempfunden. Unter dem Motto »Putzen wie die Großen« bindet Vileda ausgesuchte Qualitätsprodukte aus den Bereichen Bodenreinigung und Wäschepflege in die Spielzeugkollektion mit ein. […] [20.2]

Die *Bosch mini Work-Station* ist mit vielen, nachgeahmten Werkzeugen ausgestattet. Logische Abfolgen und taktile Fähigkeiten können durch drehen, schrauben oder kurbeln der mitgelieferten Teile erlernt werden. In einer Werbesendung auf der Internetseite der *Theo Klein GmbH* demonstrieren zwei Jungen, wie gut es sich an der Station spielen und lernen lässt. ■

67 Verpackung ›Bosch mini Work-Station‹
67 Packaging ›Bosch mini Work-Station‹
68 Detail ›Bosch mini Work-Station‹
68 Detail ›Bosch mini Work-Station‹

68

Vileda brings a breath of fresh air into every day play with its special edition for children. [...] The Vileda brand is synony-mous with future-proof deve-lopment and strong quality products. Growing potential to experience things, goes hand in hand with a forward-loo-king product and brand strategy. And this is our strategic approach: Children are our clients of the future. The posi-tive brand image which we can already instil early on in the form of high qua-lity toys suitable for children makes a big contribution to imparting emotions to the Vileda brand and our products. [...] With a love of detail, the Vileda miniatures replicate all of the main product features of the originals. True to the slogan »Cleaning like the adults« Vileda incorporates selected quality products used for cleaning floors and washing into the toy collection. [...] [20.2]

The *Bosch mini work-station* is equipped with many replica tools. Logical sequences and tactile skills can be learned by turning, twisting or rotating the parts of the work station. In an advertisement on the *Theo Klein GmbH* website, two boys demonstrate what fun it is to play and learn using the station. ■

21 Classic Alox; Victorinox Handyman; Victorinox

◂ 62

69

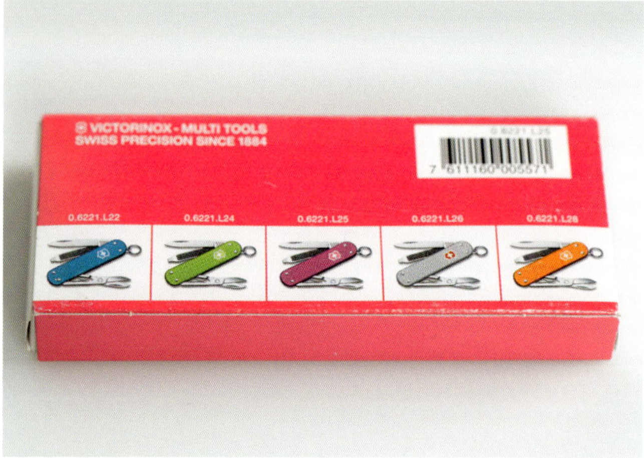

70

Der Vergleich der beiden Taschenmesser ist ein Beispiel für Komplexität und Größe als genderspezifische Unterscheidungsmerkmale. Das als Schlüsselanhänger konzipierte Taschenmesser *Classic Alox* ist neben einer einfachen Klinge lediglich mit einer Minischere sowie einer Nagelfeile ausgestattet. Die geschlechtliche Ausrichtung wird durch die Farbgebung der Produktreihe, die sich an eine jüngere weibliche Zielgruppe richtet, unterstützt.

Im Gegensatz dazu erscheint das *Handyman* als voll funktionsfähiges Multitoolwerkzeug, das seinem Nutzer in allen Lebenslagen hilfreich zur Seite steht. Die Farbgebung und Oberflächenbearbeitung ist klassisch und ohne Wertung.

Die neue, limitierte Classic-Serie – für alle, die gerne Akzente setzen. Das traditionelle Classic in neuen leuchtenden Farben. Als hübsches kleines Geschenk, als praktischer Schlüsselanhänger oder wegen der tollen Farbe – das »Kleine« passt überall. [21.1]

Große Klinge, kleine Klinge, Korkenzieher, Dosenöffner mit kleinem Schraubendreher (auch Kreuzschrauben), Kapselheber mit Schraubendreher, Drahtabisolierer, Stech-Bohr-Nähahle, Ring Inox, Pinzette, Zahnstocher, Schere, Mehrzweckhalter (Paketträger), Holzsäge, Holzmeißel, Nagelfeile mit Metallfeile, Nagelreiniger, Metallsäge, feiner Schraubendreher, Kombizange mit Drahtschneider [21.2] ■

69 Farbvarianten ›Victorinox Clasic Alox‹
69 Colour varations ›Victorinox Clasic Alox‹
70 Verpackung ›Victorinox Clasic Alox‹
70 Packaging ›Victorinox Clasic Alox‹
71 Detail ›Victorinox Handyman‹
71 Detail ›Victorinox Handyman‹

The two penknives are an example of how complexity and size are used as gender-specific differentiating characteristics. The *Classic Alox* designed as a key ring only has a simple blade, a pair of mini scissors and a nail file. The sex which this product is targeted at is accentuated by the colour scheme used in the product line, which is aimed at the younger female target group.

The *Handyman* by way of contrast appears as a fully functional multi-tool kit which provides its user with helpful support in any situation. The colouring and the surface finish is classic and completely neutral.

The new limited edition Classic series – for all those who like to leave a mark. The traditional Classic in new luminous colours. As a nice little present, as a practical key ring, or just because of the fantastic colours – the »little one« is perfect everywhere. [21.1]

Large blade, small blade, corkscrew, tin opener with a small screwdriver (also Phillips screws), bottle opener with screw-driver, wire stripper, reamer-punch & sewing eye, key ring, tweezers, toothpick, scissors, multi-purpose hook (parcel carrier), wood saw, wood chisel, nail file with metal file, nail cleaner, metal saw, fine screw-driver, combo-pliers with wire cutter. [21.2] ■

71

22 Air Vignon; menu Leverpull LM-200; Screwpull

◂ 64

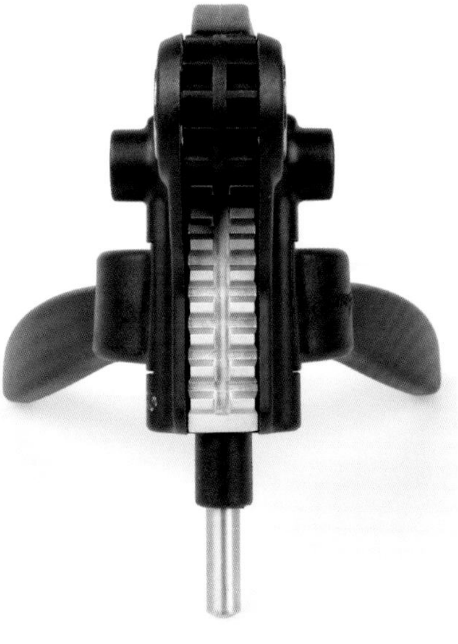

Die Gegenüberstellung der beiden Korkenzieher zeigt eine Unterscheidung durch Komplexität und Formgebung. Neben der Gestaltung wird die genderspezifische Ausrichtung der Objekte auch in den Produktbeschreibungen deutlich.

Der elegante und sanft geschwungene Körper des *Air Vignon* sowie seine Funktionsweise und Namensgebung positionieren das Produkt in einem weiblichen Feld.

Im Gegensatz dazu stellt der *Leverpull LM-200* mit seiner technischen und robusten Gestaltung das männliche Pendant dar. Der Korkenzieher wird hier als Werkzeug und Maschine gezeigt, welche durch Schnelligkeit, Effizienz, Wertigkeit und Einzigartigkeit überzeugt. ▸

A comparison of the two corkscrews reveals differences in their complexity and shape. In addition to the design, the gender-specific orientation of the articles is also clearly shown in the product descriptions.

The elegant and gently curving body of the *Air Vignon*, as well as its modus operandi and name, position the product in the feminine segment.

By way of contrast, the *Leverpull LM-200* with its technical and rugged design is male orientated. The corkscrew is here presented as a tool and a machine which impresses by way of its speed, efficiency, quality and uniqueness. ►

73

72 Detail ›Leverpull LM-200‹
72 Detail ›Leverpull LM-200‹
73 ›Leverpull LM-200‹
73 ›Leverpull LM-200‹

74

Fast sieht es aus wie Zauberei, doch es ist wahr: Mit nur ein wenig Luft befördert der Luftkorkenzieher den Korken aus der Flasche. Man steckt die Nadel durch den Korken und schon nach cirka fünfzehn Pumpvorgängen gleitet der Korken leicht und elegant aus der Flasche. Ein wirkungsvolles System in attraktiver Form vom international ausgezeichneten Designer Jakob Wagner. Der Korkenzieher eignet sich nur für echte Korken. [22.1]

Oft kopiert und nie erreicht: Der original Leverpull von Screwpull. Design-Klassiker und mechanisches Wunderwerk. Die verbesserte Ausführung ermöglicht es bis zu 2000 Flaschen zu öffnen, ohne die Spirale zu wechseln. Weltberühmt für ihre einzigartige Funktionsweise: Die Leverpull-Modelle von Screwpull gehören zu den schnellsten Korkenziehern der Welt. Tausendfach getestet und bewährt. Die Excalibur-Beschichtung der Spirale lässt bis zu 2000 Öffnungen ohne Probleme zu. Danach kann die Spirale einfach ausgetauscht werden.Und durch die Verwendung einer verkürzten Spirale bei allen Leverpulls wird das Durchstoßen von langen Korken ab 54 mm verhindert. [...] Durch die kürzere Spirale verringert sich die Möglichkeit, dass Korkkrümel in den Wein gelangen. [22.2] ■

74 Verpackung ›Air Vignon‹

74 Packaging ›Air Vignon‹

75 Verpackung ›Leverpull LM-200‹

75 Packaging ›Leverpull LM-200‹

It almost looks like magic, but it's true: this air corkscrew only requires a small amount of air to push corks out of the bottle. Just stick the needle through the cork and after around 15 pumps, the cork slides easily and elegantly out of the bottle. An effective system in an attractive design from the internationally acclaimed designer Jacob Wagner. The corkscrew is only suitable for real corks. [22.1]

Often copied but never bettered: the original Leverpull from Screwpull. A classic design and mechanical wonder. The improved design enables up to 2000 bottles to be opened without replacing the spiral. World-famous for its unique modus operandi: the Leverpull models by Screwpull are among the fastest corkscrews in the world. Tried-and-tested thousands of times. The Excalibur coating on the spiral enables the problem-free opening of up to 2000 bottles. Then simply replace the spiral. And the shorter spiral now used in all Leverpulls avoids punching through the bottom of corks longer than 54 mm [...] The shorter spiral reduces the chances of cork crumbs falling into the wine. [22.2] ■

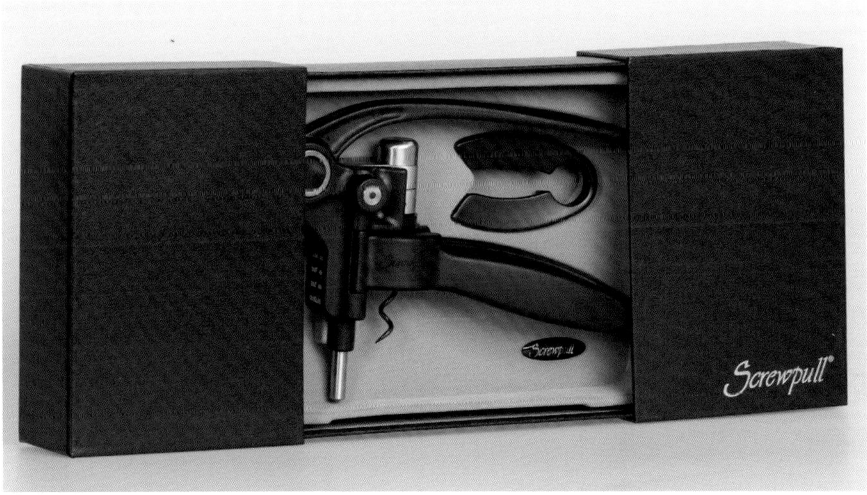

75

23 Storchschere; Wasa Solingen Arbeitsschere; Forum

◂ 66 Die Produktgegenüberstellung der *Storchschere* und der *Arbeitsschere* zeigt deutlich, wie die Größe eines Objektes den Nutzer definieren kann. Die Griffdimensionen der Stickschere sind für größere (Männer-) Hände eher ungeeignet, während Größe und Gewicht der Arbeitsschere die Benutzung für kleinere (Frauen-) Hände zu einer rasch ermüdenden Tätigkeit werden lassen.

76 Detail ›Storchschere‹
76 Detail ›Stork Scissors‹
77 Detail ›Arbeitsschere‹
77 Detail ›Working Scissors‹

76

Die Stickschere weist durch ihre Proportion und die Form sowie in Material und Materialkombination und Motivwahl eine genderspezifische Codierung auf. Das Handwerkszeug erhält Anzeichen eines dekorativen Accessoires.

Die Gestaltung der Arbeitsschere verzichtet auf eine offensichtliche geschlechtsspezifische Codierung und wirkt somit als Handwerkszeug mit technologischem Bezug aufgrund der Verarbeitung und Materialwahl. Trotz Codierung sind beide Scheren sehr funktional und mit Bezug zum jeweiligen Tätigkeitsbereich gestaltet. ■

Comparison of the *Stork scissors* and the *working scissors* clearly demonstrates how the size of an object can define the user. The size of the handles of the embroidery scissors is probably unsuitable for larger (male) hands, whilst the size and the weight of the *working scissors* will soon make their use very tiresome when used by smaller (female) hands.

77

The proportions and shape, as well as the materials and material combinations and choice of emblem used in the embroidery scissors, all reveal gender-specific coding. The hand tool takes on the appearance of a decorative accessory.

The design of the working scissors dispenses with obvious male or female specific coding so that the finish and choice of materials label this as a hand tool with a technical application. Despite the coding, both scissors are very functional and specifically designed for each area of application. ■

24 Pretty Useful Tools™ Hammer; Picard

◂ 68

Die *Pretty Usefull Tools™* werden in Deutschland unter dem Namen *Tool-Kit für Frauen* angeboten. Ursprünglich wurden die Werkzeuge von zwei Frauen aus Südafrika entwickelt, um der *langweiligen Heimwerkerarbeit mehr Stil* zu verleihen. Dies, sowie die Intention, praktikable, variationsreiche und exklusive Werkzeuge für Frauen zu gestalten, wird durch die Produkte jedoch nur teilweise erreicht. Die Tools haben einen eher unpraktischen und auf Äußerlichkeit bedachten Charakter. Die Nutzung ist mangelhaft, da die Griffe unergonomisch und die Werkzeuge teilweise sehr schwer sind. Außerdem möchte man die Produkte aus Angst, die dekorative Oberfläche zu verletzen, nicht im vollen Umfang gebrauchen. ▸

78

79

Pretty Useful Tools™ are sold in Germany under the name *Tool-Kit für Frauen*. The tools were originally developed by two women from South Africa to bring more style to boring DIY activities. However, this, as well as the intention of designing practical, varied and exclusive tools for women, is only partially satisfied by the products. The tools tend to have a rather unpractical character focused on looks alone. They function unsatisfactorily because the handles are un-ergonomic and the tools are very heavy in some cases. Moreover, one hesitates to use the products wholeheartedly for their intended purpose for fear of damaging the decorative surface.

The intended clientele for these products – women – are targeted more by the external appearance than by any real improvement in their function. This superficiality is reflected in the German product description of an internet mail order company which works with trivialisation and clichés: ▸

78 ›Pretty Usefull Tools™‹
78 ›Pretty Usefull Tools™‹
79 Detail Hammer ›Pretty Usefull Tools™‹
79 Detail hammer ›Pretty Usefull Tools™‹

Die Adressaten dieser Objekte – Frauen – werden mehr durch das oberflächliche Erscheinungsbild als über eine wirkliche Verbesserung der Benutzung angesprochen. Diese Oberflächlichkeit spiegelt sich in der deutschen Produktbeschreibung eines Internetversandes wieder, die mit Verniedlichungen und Klischees arbeitet:

Einen Nagel in die Wand hauen, das machen wir Mädels normalerweise per Augenaufschlag: »Schatz, könntest Du mal bitte …?« Aber wenn Hammer, Zange und Schraubenzieher so poppig bunt daherkommen wie bei diesem Tool-Kit, nehmen wir die Sache mit Freuden selbst in die Hand. Zum Schraubenzieher gehören vier verschiedene Aufsätze, und sollte uns das nicht reichen, schrauben wir einfach den Griff des Hammers ab in ihm befinden sich vier weitere Schraubenzieherchen. Wie heißt es doch so schön: Selbst ist die Frau! Und das Wimpernklimpern können wir von jetzt an für andere Dinge einsetzen.[24.1]

80

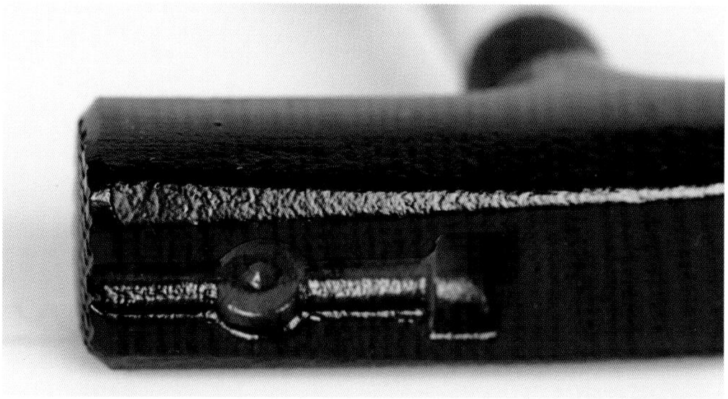

Der Hammer der Firma Picard in Baumarktqualität wirkt durch sein hohes Gewicht und seine Größe durchsetzungsstark. Die gewohnte Oberflächenqualität und Verarbeitung schaffen Vertrauen. Die strukturierte Griffzone sowie eine gute Gewichtsbalance beeinflussen die ergonomische Handhabung positiv. ■

81

Banging a nail into the wall... We girls usually do this by fluttering our eyelashes: »Darling, could you come here for a minute...?« But when hammers, pliers and screwdrivers are as brightly coloured as in this tool kit, we like to get to grips with things ourselves. The screwdriver includes four different tips, and if that is not enough, we simply screw off the handle of the hammer to find four more screw driver tips. You know what I mean: A woman's place is in the home improvement centre! And we can reserve our eyelash-fluttering for other occasion.[24.1]

The hammer made by *Picard* in DIY market quality makes a forceful impression because of its heaviness and its size. The familiar quality of the surfaces and the finishing give them a reliable look. The structured gripping zone and the well balanced feel have a positive influence on the ergonomic handling. ■

80 Detail Hammerkopf ›Picard‹

80 Detail hammer head ›Picard‹

81 Detail Hammergriff ›Picard‹

81 Detail hammer grip ›Picard‹

25 sixx; fem Media
DMAX; Discovery Communications Deutschland

◂ 70

82

Sixx – Die neue Senderin. Sixx ist emotional, selbstbewusst, lebhaft und anspruchsvoll. Sixx hat eben die moderne Frau im Visier. [...]

Da holen die Frauen schon mal den Prosecco aus dem Kühlschrank, denn jetzt kann man Sixx gemütlich genießen, wann immer Frau es möchte. [...]

Sixx ist eben anders, Sixx ist neu und weiblich – eben die neue Senderin im deutschen TV. Ab demnächst kann man auch die erste Topmodel Staffel im Serien-Channel Sixx schauen und zusammen mit den Freundinnen und ein paar Flaschen Sekt kichernd auf der Couch genießen. Was ist besser, als ein lustiger Frauenabend mit der neuen Senderin Sixx? Nichts. [...] [25.1]

Die Webseite des Frauensenders *sixx* bietet Informationen zu Themen wie *Stars, Style und Lifestyle, Romantische Date Locations, Suri Cruise hat ihre eigene Kreditkarte und Designer-Brautkleider von Jason Wu.*

Der Männersender *DMAX* wirbt auf seiner Internetseite mit dem Slogan *Fernsehen für die tollsten Menschen der Welt: Männer.* Programmtitel wie *DMAX Speed, Die Gebrauchtwagen-Profis, Zerstört in Sekunden, Holzfäller extrem* oder *Das Hausbau-Kommando* geben bereits einen Einblick in die *DMAX* Welten Adrenalin, Lifestyle, Motor, Typen. [25.2] ■

82 Werbung ›sixx‹

82 Advertisment ›sixx‹

83 ›Ausgesetzt in der Wildnis‹ Copyright DMAX

83 ›Exposed to the wilderness ‹ Copyright DMAX

Sixx – The new channel. Sixx is emotional, self-confident, lively and sophisticated. No wonder, Sixx is aimed at modern women. [...]

So what if women get a bottle of Prosecco out of the fridge occasionally, – they have a right to enjoy Sixx in comfort, whenever they want. [...]

Sixx happens to be different, Sixx is new and feminine – the new channel on German TV actually. Soon, you will also be able to see the first top model series in the Sixx series channel, and enjoy a giggle on the sofa with your girlfriends and a couple of bottles of sparkling wine. What is better than an amusing evening with the girls and the new Sixx channel? Nothing. [...] [25.1]

The website of the sixx women's channel provides information on topics such as *stars, style and lifestyle, romantic date locations, Suri Cruise has her own credit card and Designer wedding dresses from Jason Wu.*

The men's channel *DMAX* advertises on its website with the slogan *Television for the most fantastic people in the world: men.* The programmes with names like *DMAX Speed, The second hand car professionals, Destroyed in seconds, Extreme lumberjacks, or The house building squad,* already give an insight into the *DMAX* worlds of adrenalin, lifestyle, motor, guys. [25.2] ■

83

26 Kadett 912 Ideal Westernland; J.G.Schrödel VIP Ideal Westernland; J.G.Schrödel

◄ 72

84

85

Die beiden Spielzeugpistolen der Serie IDEAL unterscheiden sich nicht nur in ihrer äußeren Erscheinung, sondern auch in der Haptik und Funktionalität. Das rosafarbene Modell *Kadett 912* ist aus Kunststoff mit eingeschlossenen Glitzerpartikeln gefertigt. Der aus einem Guss produzierte Revolver hat auf Grund der Materialwahl ein geringes Gewicht, was das Produkt unecht wirken lässt. Die eingeschränkte Funktionsfähigkeit der unbeweglichen Trommel wertet das Produkt zusätzlich ab.

Das Modell *VIP* wurde bis auf den Griff aus Metall verarbeitet. Das dadurch entstehende Gewicht lässt das Produkt hochwertiger erscheinen. Die Revolvertrommel dreht sich beim Betätigen des Abzugs. Der Holz imitierende Kunststoffgriff liegt gut in der Hand und wird seitlich mit einem kleinen, golden geprägten Signet aufgewertet. ■

84 Detail Technik ›Kadett 912‹
84 Detail technic ›Kadett 912‹
85 Detail Holster ›Kadett 912‹
85 Detail holster ›Kadett 912‹
86 Detail Technik ›VIP‹
86 Detail technic ›VIP‹
87 Detail Material ›VIP‹
87 Detail material ›VIP‹

86

87

The new toy guns in the *IDEAL* series not only look different from one another, they also have different haptics and functionalities. The pink model *Kadett 912* is made of plastic with embedded glitter. The one-piece injection-moulded revolver is light because of the material it was made of, which also makes it feel unauthentic. The product is further diminished by the restricted functionality of the rigid drum.

The VIP model is made of metal with the exception of the grip. It therefore weighs more and this makes the product appear more valuable. The revolver drum rotates when the trigger is pulled. The wood-imitation plastic handle can be gripped comfortably, and the quality is enhanced further by the small golden embossed signet on the side. ■

27 Ergomat; Fiskars

◂ 76

88

Die heute in den USA ansässige Firma *Fiskars* wurde bereits 1649 im gleichnamigen finnischen Dorf gegründet. *Fiskars* legt Wert auf ergonomische und sichere Produkte mit einer langen Lebensdauer. *Profis und Endverbraucher finden unsere Werkzeuge inspirierend und ideal geeignet für anspruchsvolle Aufgaben.* Die komfortablen, qualitativ hochwertigen Produkte werden für Menschen jeden Alters und verschiedenster Lifestyles entwickelt. [27.1]

Zu den bekanntesten Produkten zählt die Schere *Ergomat* mit ihren orangefarbenen, ergonomisch geformten Griffen, die bereits in den 1960er Jahren erstmals gestaltet und produziert wurde. Die Schere *Ergomat* ist geschlechtsneutral und nutzerorientiert gestaltet und zählt somit zum Bereich des Universal Design.

Seit Jahrzehnten die Ikone für ausgezeichnetes ergonomisches Design. Diese Schere passt perfekt und ist optimal zu bedienen und zu führen. Ein ideales Handwerkzeug für Büro, Haushalt und Werkstatt. [27.2] ■

88 Ergonomischer Griff ›Ergomat‹

88 Ergonomic handles ›Ergomat‹

89 Detail ›Ergomat‹

89 Detail ›Ergomat‹

Fiskars, currently headquartered in the USA, was originally founded in 1649 in the eponymous Finnish village. *Fiskars* prioritises ergonomic and safe products with long service lives. *Professionals and end users find our tools inspiring and perfectly suited to challenging tasks.* The comfortable, high quality products are developed for people of all ages and different lifestyles. [27.1]

The best-known products include the *Ergomat* scissors with their orange-coloured, ergonomically shaped handles, which were first designed and produced in the 1960s.

The *Ergomat* scissors has a gender non-specific and user-oriented design, and is therefore placed in the Universal Design category.

An icon of superb ergonomic design for decades. These scissors fit perfectly and can be optimally operated and guided. The ideal hand tool for office, home and workshop. [27.2] ■

89

28 2911 Softouch®; Fiskars

Die heute in den USA ansässige Firma *Fiskars* wurde be-
◂ 78 reits 1649 im gleichnamigen finnischen Dorf gegründet. *Fiskars* legt Wert auf ergonomische und sichere Produkte mit einer langen Lebensdauer. *Profis und Endverbraucher finden unsere Werkzeuge inspirierend und ideal geeignet für anspruchsvolle Aufgaben.* Die komfortablen, qualitativ hochwertigen Produkte werden für Menschen jeden Alters und verschiedenster Lifestyles entwickelt. [28.1]

Die Schere *2911 Softtouch* ist geschlechtsneutral und nutzerorientiert gestaltet und zählt somit zum Bereich des Universal Design.

Fiskars Komfort Schere für Rechts- und Linkshänder. Der spezielle Winkel des Griffes zur Klinge ermöglicht ermüdungsfreies Schneiden auch für Menschen mit kräftigen Fingern. Softtouch® Griffe mit federunterstützter Öffnung für einfachen Gebrauch, Komfort und Kontrolle. Einfach bedienbarer Verschluss für mehr Sicherheit. [28.2] ■

90

Fiskars, currently headquartered in the USA, was originally founded in 1649 in the eponymous Finnish village. *Fiskars* prioritises ergonomic and safe products with long service lives. *Professional and end users find our tools inspiring and perfectly suited to challenging tasks.* The comfortable, high quality products are developed for people of all ages and different lifestyles. [28.1]

The *2911 Softtouch* scissors have a gender-neutral and user-oriented design, and therefore fall into the Universal Design category.

Fiskars' comfort scissors for right-handed and left-handed users. The special angle between the handle and the blade make cutting non-tiring, even for people with large fingers. Softtouch® handles with spring-supported opening for easy use, comfort and control. Easy catch operation for extra safety. [28.2] ■

91

90 Ergonomischer Griff ›2911 Softtouch‹

90 Ergonomic handles ›2911 Softtouch‹

91 Detail ›2911 Softtouch‹

91 Detail ›2911 Softtouch‹

29 IXO; Bosch

◂ 80

92

Der Akku-Schrauber *IXO* aus der Heimwerkerserie der Firma *Bosch* ist auf die Bedürfnisse einer breiten Nutzerschicht abgestimmt. Durch das geringe Gewicht, eine handliche Größe und die gefällige, nicht ausgesprochen techniklastige Gestaltung entfernt sich das Werkzeug aus der männerdominierten Heimwerkerwelt und wird auch für *Frauen und heimwerkende Einsteiger* interessant. Die Nutzbarkeit ist im Vergleich zu ähnlichen Produkten des Wettbewerbs gut und leicht verständlich, bleibt jedoch durch die geringe Leistung des Gerätes auf einfache Heimwerkertätigkeiten beschränkt.

[...] Ein formschönes Produktdesign, die handliche kompakte Form, geringes Gewicht, einfache Bedienbarkeit und originelle Verpackungen wie der Akkuschrauber IXO aus der Geschenkdose erschließen neue Zielgruppen: Frauen und eine junge Generation von heimwerkenden Einsteigern. [29.1] ■

93

The *IXO* rechargeable screwdriver from *Bosch's* DIY series is tailored to the needs of a wide range of users. Its low weight, handy size, and attractive and not overstated technology-oriented design, distances the tool from the male-dominated DIY world, and makes it interesting for *women and DIY newcomers* as well. Its practicality is good compared to similar products by other companies, and easy to understand – but is restricted to simple DIY jobs because of the low performance of the device.

[...] An attractive product design, the handy compact shape, low weight, simple operation and eye-catching packaging – such as the IXO rechargeable screwdriver in the gift box – open up new target groups: women and a young generation of DIY newcomers. [29.1] ■

92 Box für ›IXO‹ mit Anmutung einer Geschenkdose

92 Case for ›IXO‹ resembling a gift box

93 Ladestation ›IXO‹

93 Loading unit ›IXO‹

30 Ergo; Bahco

◄ 82

Die Produktlinie der Schraubendreher *ERGO* von *Bahco* ist ein Beispiel für eine geschlechtsneutrale universale Gestaltung, die sich an den Bedürfnissen des Nutzers orientiert, ohne sich oberflächlicher und scheinbar notwendiger Gestaltungselemente zu bedienen.

Das Büro *Ergonomidesign* aus Schweden begann bereits 1978 mit der Entwicklung von ergonomischen Schraubendrehern für die Firma *Bahco*. Die Produkte wurden in Zusammenarbeit mit Handwerkern entwickelt und getestet. Die besondere Gestaltung des Griffes ermöglicht ein beidhändiges Bedienen des Schraubendrehers, so dass eine optimale Kraftumsetzung erfolgen kann. ■

94

94 Detail Griffe Schraubendreher ›ERGO‹

94 Detail handles screwdriver ›ERGO‹

95 Kennzeichnung Schraubendreher ›ERGO‹

95 Marking screwdriver ›ERGO‹

95

The *ERGO* screwdriver product line from *Bahco* is an example of non-gender-specific Universal Design which is oriented to the needs of the users without adopting superficial and apparently necessary design elements.

Ergonomidesign in Sweden began to develop ergonomic screwdrivers for *Bahco* back in 1978. The products are developed and tested together with craftsmen. The specially designed handle enables the screwdriver to be operated with two hands to optimise the force applied to the screwdriver. ■

31 WMF 1; WMF

Die *WMF 1* erhielt aufgrund ihres
◂ 84 selbsterklärenden Designs und der leichten Bedienbarkeit nur durch eine einzige Funktionstaste, das *Qualitätszeichen Universal Design*.

Das *Qualitätszeichen Universal Design* wird vom *Design Zentrum Berlin*, dem *Rat für Formgebung* und dem *TÜV NORD CERT* verliehen. Es wird an Produkte vergeben, die den Kriterien des Universal Design entsprechen: also gut handhabbar, technisch sicher und generationenübergreifend attraktiv sind. Durch diese Auszeichnung erhält der Verbraucher Gewissheit, dass ein entsprechendes Produkt technisch sicher, intuitiv zu nutzen und darüber hinaus gut gestaltet ist. Außerdem werden die Produkte technisch auf Sicherheit, Materialeigenschaften, Kräfte und Normen überprüft.

Die weltweit kleinste Kaffeepadmaschine WMF 1 können Sie nahezu überall dort verwenden, wo Sie eine Tasse frisch gebrühten Kaffee genießen möchten.In weniger als einer Minute bereitet Ihre WMF 1 eine Tasse duftend frischen Kaffee zu. Der Wassertank fasst so viel, wie Sie für eine Tasse Kaffee brauchen. Sie brühen also immer mit frischem Wasser. Legen Sie in den Padhalter ein handelsübliches Kaffeepad Ihrer Wahl. Einfach Knopf drücken und in weniger als einer Minute ist Ihre Tasse duftend frischer Kaffee fertig.

Die WMF 1 ist Kaffeepadmaschine und Design-Objekt zugleich. Sie überrascht durch ihre außergewöhnliche Formgebung und die integrierte Tasse und macht damit jede Kaffeepause zu einem besonderen Ereignis. [31.1] ■

96

96 ›WMF 1‹ funktionsbereit
96 ›WMF 1‹ operational
97 ›WMF 1‹ mit Padhalter
97 ›WMF 1‹ with pad holder
98 ›WMF 1‹
98 ›WMF 1‹

The *WMF 1* is awarded the *Universal Design Quality Label* because of its self-explanatory design and easy operation with only one function button.

The *Universal Design Quality Label* is awarded by the *Design Zentrum Berlin*, the *German Design Council*, and *TÜV NORD CERT*. It is awarded to products which satisfy the Universal Design criteria: in other words, easy to use, technically safe and attractive for all generations. This award guarantees users that the product is technically safe, can be used intuitively, and is also well designed. The products are also examined technically to check their safety, material properties, forces and standards.

The WMF 1 is the world's smallest coffee pad machine and can therefore be used almost everywhere – wherever you want to enjoy a cup of freshly brewed coffee. Your WMF 1 prepares a cup of fresh aromatic coffee in less than one minute. The water tank holds just the right amount for one cup of coffee. You therefore always brew coffee with fresh water. Just put your favourite brand of coffee pad in the pad holder. Simply press the button, and your cup of fresh aromatic coffee is ready in less than a minute.

The WMF 1 is a coffee pad machine and a design article in one. It catches the eye with its unusual shape and the integrated cup, making every coffee break a special event. [31.1] ■

97

98

32 Cookware; Royal vkb

Das Kochgeschirr von Royal vkb ◂ 86 *besteht aus schönen und praktischen Kochtöpfen, die international mehrfach ausgezeichnet wurden. Mithilfe des intelligenten Verschlusssystems können Sie durch einfaches Drehen des Deckels heißes Wasser abgießen, ohne sich zu verbrennen. Alle Kochtöpfe haben eine Messskala und sind ineinander stapelbar. Dadurch sparen Sie Platz in Ihren Küchenschränken. Die matten Griffe der Edelstahltöpfe sind hitzebeständig und die Töpfe gibt es in unterschiedlichen Größen und Farben. Das Kochgeschirr ist für alle Herdarten einschließlich Induktionsherde geeignet, backofentauglich und spülmaschinengeeignet.* [32.1]

Der funktionale Charakter der Topfserie wird gestalterisch konsequent umgesetzt. Die Nutzung der Töpfe wird Menschen mit unterschiedlichen Ansprüchen gleichermaßen gerecht – so ist beispielsweise das einhändige Abgießen des Wassers durch einfaches Verschließen des Deckels möglich. Eine geschlechtliche Zuordnung gibt es nicht. ■

99

100

Royal vkb's Cookware comprises beautiful but also very practical internationally awarded cooking pans. The smart locking system allows you to twist the lid and strain boiling water without any difficulty. All pans have a measuring scale and are stackable, saving a lot of space. The stainless steel pans with matt handles are fire-resistant, available in different sizes and trendy colours. The Cookware is suitable for all heat sources including induction, the oven and the dishwasher. [32.1]

The functional character of this pan series is uncompromisingly implemented. The pans can be used by people with a range of expectations – for instance, water can be poured out with one hand by simply locking the lid. There is no gender-specific characterisation. ■

99 ›Cookware‹ Ausgießer

99 ›Cookware‹ spout

100 ›Cookware‹ Messskala

100 ›Cookware‹ measuring scale

33 brand eins; brand eins Verlag

101

Was ist die Philosophie von brand eins?
◂ 88 *brand eins ist das Wirtschaftsmagazin, das nach den Hintergründen sucht und nach den Zusammenhängen. Wir nehmen scheinbar Vertrautes auseinander und setzen es neu zusammen, wir kreuzen Wirtschaft mit Kultur und Gesellschaft. Unser Angebot ist der Perspektivwechsel – denn neue Sichtweisen sind entscheidend für eine Wirtschaft, in der Kreativität und Wissen die wichtigsten Produktivfaktoren sind. brand eins beschreibt den momentanen Wandel in Wirtschaft und Gesellschaft, den Übergang vom Informations- zum Wissenszeitalter. brand eins zeigt die Bruchstellen, die sich dabei ergeben und liefert Vorlagen, Ideen und Konzepte für alle, die diesen Wandel aktiv vorantreiben oder von ihm berührt werden.* [33.1]

Sowohl der Titel als auch das gesamte Layout des monatlich erscheinenden Magazins sind großzügig und übersichtlich gestaltet und geben einen geschlechtsneutralen Eindruck. Die verwendete Serifenschrift verleiht der Zeitschrift einen klassischen und seriösen Charakter, ohne konservativ zu wirken. Die in Pixelschrift gesetzten Kopfzeilen stehen dazu in wirkungsvollem, technisch anmutendem Kontrast. Ganzseitige Bilder sowie Grafiken führen zu einem spannungsvollen Wechsel von Schrift und Bild. Die Themen werden teilweise in Form von Personenportraits aus verschiedensten Lebensbereichen beleuchtet, was die Auseinandersetzung und Identifikation mit Wirtschaftsthemen erleichtert und den Leserkreis erweitert. ■

What is the philosophy of brand eins? brand eins is a business magazine which searches for the background facts and the relationships. We dissect things that appear to be familiar and put them back together again in a new way, we combine business with culture and society. What we offer is a change of perspective – because new points of view are crucial for an economy in which creativity and know-how are the most important productivity factors. brand eins describes the current changes affecting business and society, the transition from the information to the knowledge age. brand eins highlights the tensions this causes and provides criteria, ideas and concepts for everyone wanting to actively push this transition forward or affected by this change. [33.1]

Both the title and the whole layout of this monthly magazine are designed lavishly and clearly, and create a gender-neutral impression. The serif font gives the magazine a classic and serious character without appearing conservative. The modern high-tech look of the pixel font used for the headlines effectively stands out in contrast to the serif typeface. Full page photos and graphics create an exciting juxtaposition of copy and visuals. The issues are analysed in part in the form of portraits of people from a whole range of fields – this simplifies the analysis of and identification with business issues, and enlarges the circle of readers. ■

101 Layout Seiten 88/89 ›brand eins‹ Heft 05 2011
101 Layout pages 88/89 ›brand eins‹ issue 05 2011
102 Layout Seiten 120/121 ›brand eins‹ Heft 05 2011
102 Layout pages 120/121 ›brand eins‹ issue 05 2011

„Nur gemocht zu werden reicht nicht"

102

34 Rollbag; Authentics

◂ 90

Die Gestaltung und Verwendung der Kulturtasche *Rollbag* sind geschlechtsunabhängig. Die einfache und intuitive Bedienung, sowie die große Farbpalette sind auf eine breite Nutzerschicht ausgelegt.

Ob für Wäsche beim Sport oder auf Reisen, für Kabel und Adapter oder als klassische Kulturtasche, die wasserabweisenden ROLLBAGs sind für viele Dinge gut. Mit dem intelligenten Rollverschluss passen sie sich ihrem Inhalt an. Dadurch bieten sie so viel Volumen wie nötig und sind gleichzeitig so handlich wie möglich. Robust und leicht zu reinigen. In drei Größen und vielen Farben erhältlich. Aus beschichtetem Polyestergewebe oder beschichteter Microfaser. Die kleine Rollbag bietet genügend Platz für Kabel, Kosmetika oder Schminkutensilien. Auch als robuste Alternative zur Butterbrotstüte geeignet.

Die mittelgroße Rollbag eignet sich für Wäsche – auch feuchte wie etwa Schwimmzeug oder als klassischer Kulturbeutel. Auf Reisen ist besonders praktisch, dass man sie zuerst für die saubere und später für die Schmutzwäsche benutzen kann. Die große Rollbag nimmt alles auf, was auch in die kleineren Varianten passt. Sie eignet sich darüber hinaus auch als dicht schließende Schuhtasche zum Beispiel für den Sport. [34.1] ■

103

103 Seitenansicht ›Rollbag‹
103 Side view ›Rollbag‹

The design and use of the *Rollbag* toiletries bag is non-gender-specific. The simple and intuitive operation, as well as the large range of colours, is designed for a broad spectrum of users.

The water-repellent ROLLBAGs are good for many things, such as sports kit, or clothes when travelling, for cables and adapters, or as a classic toiletries bag. With the intelligent roll-closer, they adapt to their contents. They therefore have as much volume as required, at the same time as being as compact as possible. Rugged and easy to clean. Available in three sizes and a range of colours. Made of coated polyester fabric or coated microfiber material. The small rollbag has enough room for cables, cosmetics or make-up utensils. It is also suitable as a rugged alternative to a sandwich bag.

The medium-sized rollbag is suitable for clothes – even wet clothes such as swimming things – or as a classic toiletries bag. A particularly practical aspect when travelling is that you can first use it for your clean clothes and later for your dirty clothes. The large rollbag accommodates everything that also fits in the smaller bags. It is also suitable as a tightly sealable shoe bag for sports shoes for instance. [34.1] ■

35 Funnel; Normann Copenhagen

104

◂ 92

Der faltbare Küchentrichter ist funktional und praktisch, denn seine ganze Größe entfaltet er erst bei der Benutzung. Das moderne Material ist außerordentlich form- und temperaturbeständig und so robust, dass es für ungezählte Vorgänge des Auseinander- und Zusammenfaltens geeignet ist. Der dänische Designer Boje und Normann Copenhagen haben, in enger Zusammenarbeit mit Materialentwicklern und Ingenieuren, das innovative weiche Material für die hohen Ansprüche in der modernen Küche entwickelt. Immer in Form, stellen auch große Hitze und Kälte kein Problem für diesen Küchentrichter dar, der sich bei Nichtbenutzung in eine flache Scheibe verwandelt und überall platzsparend verstaut werden kann. ›Funnel‹ ist ein leicht zu reinigender – weil spülmaschinengeeigneter – Küchentrichter, der funktional, aber auch ästhetische Akzente in jeder modernen Küche setzt. [35.1]

Die intuitive und einfache Handhabung, sowie die geschlechtsneutrale und archetypische Gestaltung machen den flexiblen Haushaltstrichter zu einem universellen Produkt, dessen Nutzer nicht auf einen bestimmten Kreis beschränkt sind. ■

104-106 Faltung ›Funnel‹

104-106 Folding ›Funnel‹

105

106

The collapsible kitchen funnel is a simple, functional and practical object, which only takes up space when it is in use. The material is flexible enough to be folded again and again, hard enough to keep its shape and durable enough to resist cold as well as hot temperatures. Boje met Normann Copenhagen, and in a co-operation with engineers all parties developed this special soft material – flexible enough to fold together, hard enough to keep its shape and durable enough to take cold and heat as well as the repetition of folding the funnel. The end product speaks for itself and apart from being space-saving and easy to clean the funnel is aesthetic. [35.1]

The intuitive and easy handling, as well as the gender-neutral and archetypal design, places the flexible household funnel in the Universal Product category whose users are not restricted to a specific group. ■

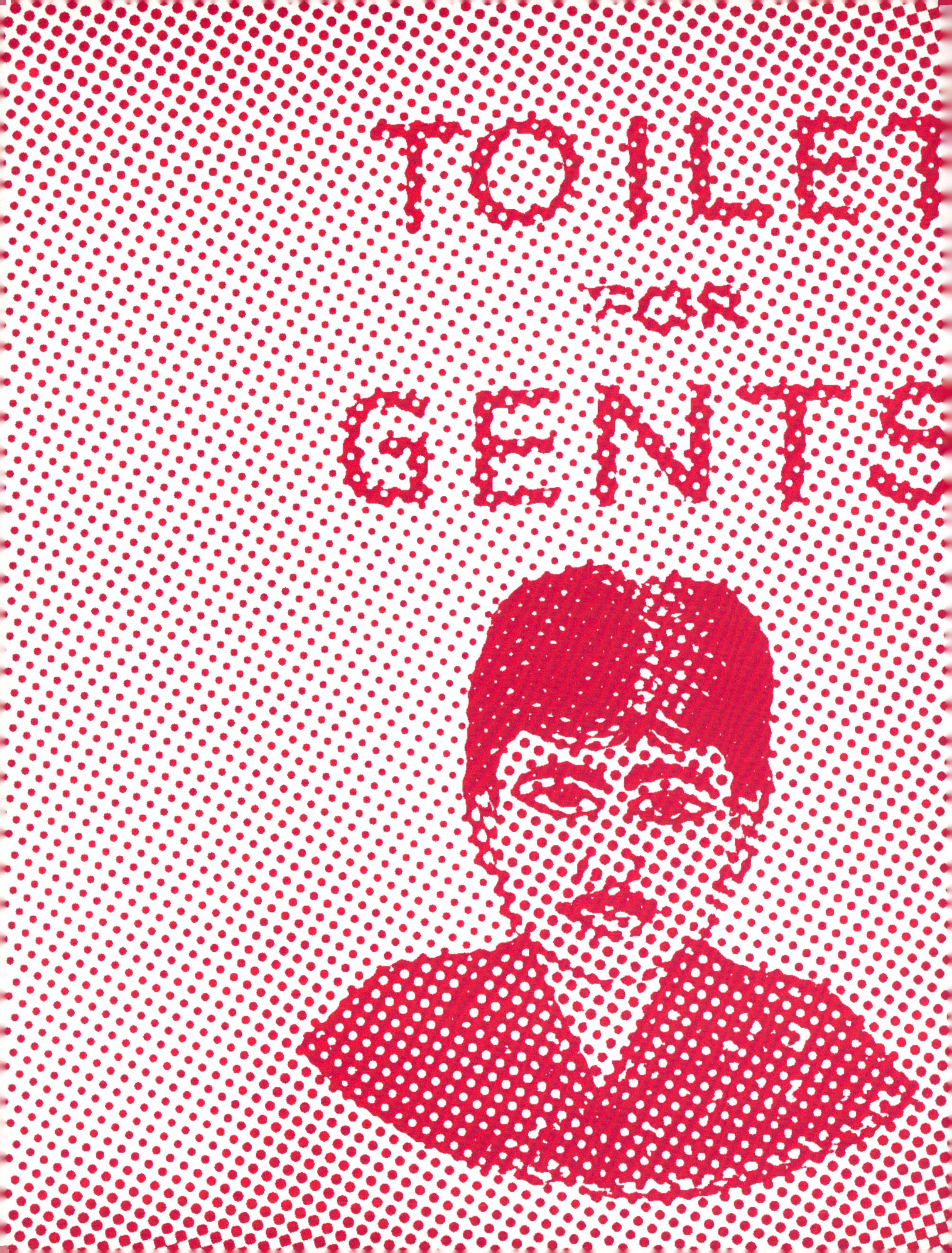
TOILET
FOR
GENTS

Raus aus dem Dschungel
Out of the jungle

Dem Dschungel entkommen...
Die Genderfalle

Von den ersten Tagen unseres Lebens – schon auf der Entbindungsstation – werden wir mit den Namensbändchen an den Handgelenken der Babies geschlechtsspezifisch codiert: zu kleinen Tarzans und Janes. Der sicherlich kleine Unterschied wird im Laufe unseres gesellschaftlichen Lebens durch unendlich viele subtile Signale zum großen Unterschied.

Neueste Forschungen aus den USA zeigen, dargestellt von der Neurobiologin Lisa Eliot in ihrem Buch *Pink Brain – Blue Brain* von 2009, dass der Unterschied zum Beginn unseres Lebens marginal ist. Weder haben weibliche Babies weniger Anlagen für mathematische Fragestellungen oder ein signifikant weniger räumliches Vorstellungsvermögen als männliche Babies einen wirklichen Mangel an sozialer Empathie besitzen. Wir bekommen es anerzogen und agieren in späteren Jahren genau nach diesen Erwartungsvorstellungen der gesellschaftlich bedingten Rollen.

Es stellt sich also die Frage, ob Designer, Marketingexperten, Verbraucherinnen und Verbraucher sich auf ewig diesem Diktat beugen müssen. Ist es denn wirklich so, dass wir Frauen nur die soften Telefone mögen? Warum haben dann ziemlich geschlechtslose Smart-Phones bei beiden Geschlechtern den gegenwärtigen Erfolg? Wollen die Männer wirklich von Kosmetik nichts wissen? Warum gibt es dann zunehmend mehr Beauty-Artikel in den Männermagazinen? ▸

Out of the jungle...
The gender trap

In the very first hours of our life on earth – as babies in a maternity ward – we become gender coded by the name bands put on our wrists: as small Tarzans and baby Janes. This probably tiny difference is turned into a very big difference during the course of our social lives by an endless chain of subtle signals.

Latest research in the US, presented by neurobiologist Lisa Eliot in her book *Pink Brain – Blue Brain* published in 2009, indicates that the differences at the start of our lives are marginal. Female babies have neither less talent for mathematics nor significantly less spatial understanding than do male babies have a genuine lack of social empathy. All suggested differences we will grow into, and in later years respond and act in a way entirely compliant with the expectations of our socially dictated roles.

The question that needs to be asked is: do designers, marketing experts, female and male users have to live with this repression forever? Is it true, that women only like soft telephones? If that's the case why are the relatively gender-less smartphones so successful today amongst both sexes? Do men really want to know nothing about cosmetics? If that's true, why are there now more beauty articles in men's mags?

Even self-aware gender behaviour is a matter of education, and determines our behaviour as consumers. Why are expensive design articles with a universal design aura sold mainly to a well educated middle class? Many of these products also demonstrate that one cannot do ▸

Auch selbstbewusstes Genderverhalten ist eine Frage der Bildung und bestimmt unser Konsumverhalten. Denn warum werden teure Designartikel mit einem universellen Designanspruch vorwiegend von der gebildeten Mittelschicht gekauft? Viele dieser Produkte beweisen auch, dass man auf Humor – das ironische Spiel mit Gender Codes oder die bewusste Überzeichnung von Klischees – nicht verzichten muss. Es kann ein Mittel für Produktdifferenzierung und Diversifizierung sein. Der selbstbewusste Umgang mit unterschiedlichen Interessen und Geschmäckern von uns Frauen und Männern ist zum Glück selbstverständlich. Doch sollten sich beide Geschlechter nicht zunehmend fragen, ob sie wirklich so »schlicht gestrickt« sind, wie man es uns oft zu vermitteln versucht?

Ganze Branchen beschneiden sich selbst in ihrer Marktfähigkeit, indem sie den Klischees folgen und die Geschlechter simplifizieren. Technisches Desinteresse bei Mädchen mit mangelndem Interesse an Legosteinen zu erklären ist einfach. Aber bietet Lego denn technisches Spielzeug mit Objekten der Begierde im Interessenkreis der Mädchen an? Wo ist der aufwendige Bausatz für einen hochmodernen Kinderwagen, der wie im wirklichen Leben eben nicht gleich in Pink daherkommen muss? Wo ist der Handmixer mit einem kleinen Elektromotor? Vielleicht fänden Jungs diesen dann auch spannend, und sie würden ihren Milchshake damit selber mixen wollen und dies nicht der Schwester überlassen? ▶

without a little humour – an ironic game with gender codes or a deliberate emphasis of a cliché sells. These are both means of product differentiation and diversification. Luckily, handling the different interests and tastes of men and women with confidence is second nature. But shouldn't both sexes be increasingly asking the question of whether they are as »simple-minded« as is often made out or not?

Whole industries lose huge slices of marketability by trotting out the old clichés and simplifications about and of the sexes. It is to simplistic to explain a girl's lack of interest in things technical by reciting a lack of interest in Lego bricks. Is Lego the right technical toy to attract girls? Can they make the kind of objects girls find desirable? Where is that sophisticated kit to assemble a state-of-the-art perambulator which, like in real life, does not have to be absolutely pink? Where is that kitchen mixer with a small electric motor? Maybe boys would actually find things like this to be of interest and enjoy mixing their own milkshakes instead of letting their sisters do it for them?

This book attempts to present not only real artefacts but also visions. The objective is not some kind of dumbing down or levelling out. It is all about the attempt to take the other sex seriously, while still being able to laugh about ourselves at the same time. It's laughing with and not at something which is the big (or little) difference. ▸

In diesem Buch wurde der Versuch gewagt, neben den realen Fundstücken auch diese Visionen zu zeigen. Denn es geht nicht um Gleichmacherei. Es geht um die Mühe, das andere Geschlecht ernst zu nehmen, ohne nicht auch über uns gegenseitig lachen zu können. Lachen und nicht Auslachen, das ist der kleine Unterschied.

Wagen wir in Zukunft auch ein »Nein« zu zuviel dümmlicher Kategorisierung, ohne dabei unseren Humor zu verlieren. ■

Katrin Hinz

In the future we should have the courage to say »No!« to many of these stupid categorisations, but without losing our sense of humour when doing so. ■

Katrin Hinz

Sprachschatz
Vocabulary

Gender | Gender 204

weiblich | female 206

männlich | male 208

biologischer Determinismus | Biological Determinism 210

androgyn/unisex | androgyny/unisex 216

Stereotype | Stereotype 218

Genderisierung | Genderisation 220

Geschlechterrollen | Gender Roles 224

Geschlechterdifferenz | Gender Difference 226

Gender Mainstreaming | Gender Mainstreaming 230

Gender Design | Gender Design 232

Gender Codes | Gender Codes 234

Genderspezifisches Design | Gender-specific Design 236

Gendersensibles Design | Gender-sensitive Design 238

Diversity | Diversity 240

Gender

Für den aus dem Englischen stammenden Begriff gender *gibt es im Deutschen bislang keine allgemein eingeführte Entsprechung; ursprünglich nur auf das grammatische Geschlecht der Substantive bezogen (in der lat. Grammatik mit Genus bezeichnet), verweist der Begriff darauf, dass Geschlechtsidentität nicht angeboren, sondern sozio-kulturell durch diskursive Zuschreibung erworben wird. Der Begriff ist somit Ausdruck der Einsicht, dass Weiblichkeit und Männlichkeit historisch-zeitgebundene Konstruktionen sind. Die Entdeckung der Geschichtlichkeit sexueller Identität gehört zu den wichtigsten Ergebnissen der noch relativ jungen* Gender Studies, *deren Genese wiederum Teil eines Perspektivenwechsels von der Ereignis- und Politikgeschichte zur Mentalitäts-, Alltags- und Kulturgeschichte ist. [...]* ■

Waltraud Wara Wende: »Gender«, in: Metzler Lexikon Gender Studies, Geschlechterforschung. Herausgegeben von Renate Kroll. Seite 141-142.

Gender

Gender *originally referred to the grammatical* gender *of a noun and is today used to indicate that sexual identity is not something innate, something one is born with but which is acquired socio-culturally by discursive attribution. The term is therefore an expression of the view that femininity and masculinity are both historical and contextual constructions. This realisation, of the historical aspect of sexual identity, is one of the key events in the relatively young interdisciplinary field of* gender studies, *which has itself evolved as part of a transforming perspective, from an event and political history through towards mentality-, daily- and cultural history. [...]* ■

Waltraud Wara Wende: »Gender«, in: Metzler Lexikon Gender Studies, Geschlechterforschung. Edited by Renate Kroll. Page 141-142.

weiblich

Jahrhundertelang wurde der Begriff der Weiblichkeit mit Attributen in Verbindung gebracht, die der vermeintlich naturgegebenen Schwäche und Gefühlsbetontheit der Frau Ausdruck verleihen: Anmut, Schönheit, Güte, Schamhaftigkeit, Schwäche, Bescheidenheit, Passivität, Affektivität (Gefühl) usw. In der Frauen- und Geschlechterforschung wird der traditionelle Begriff der Weiblichkeit als das Resultat androzentrischer Definitionsmacht verstanden.

Weiblichkeit wird, vor allem auch nach den Erkenntnissen in den Gender Studies, *nicht mehr als angeboren oder durch Anatomie, Biologie, Sexualität und Psyche determiniert erkannt, sondern als gesellschaftlich und kulturell geschaffenes Konstrukt gesehen. Der Begriff Weiblichkeit geht auf Aristoteles, d.h. seine Festlegung der ›weiblichen Natur‹ auf Irrationalität zurück. [...]* ■

Elena Kilian: »Weiblich / Weiblichkeit«, in: Metzler Lexikon Gender Studies, Geschlechterforschung. Herausgegeben von Renate Kroll. Seite 399-400.

female

For many centuries the term feminine was associated with attributes expressing the purportedly natural weakness and emotionality of women: grace, beauty, kindness, modesty, weakness, chastity, passivity, affectivity (feelings), etc. In the fields of women and gender research, the traditional term feminine is now understood as the consequence of androcentric definatory power.

Femininity is above all, according to the results of gender studies, *no longer considered to be innate or determined by anatomy, biology, sexuality or the psyche, but as societal and cultural construct. The term feminine is attributed to Aristotle, i.e. to his definition of ›feminine nature‹ as irrational. [...]* ■

Elena Kilian: »Weiblich / Weiblichkeit«, in: Metzler Lexikon Gender Studies, Geschlechterforschung. Edited by Renate Kroll. Page 399-400.

männlich

*Der Begriff männlich/Männlichkeit bedeutet »dem Mann angemessen, tapfer, mutig« (*Duden Etymologie *1989, S. 438). Diese Eigenschaften spiegeln eine Geschlechterideologie wider, die dem männlichen Geschlecht ›natürliche‹ Attribute zuordnet. Darüber hinaus werden an diesen Zuweisungen und Geschlechtereigenschaften – besonders auch im Vergleich zum Weiblichen bzw. zur Weiblichkeit – Herrschaftskonstruktionen deutlich. [...]*

Der umgangssprachliche Ausdruck Macho bezeichnet einen Mann mit einem übersteigerten Männlichkeitsgefühl (vgl. Duden Etymologie *1989, S. 431). Das aus dem Mexikanischen entlehnte Wort geht auf lat.* masculus *(männlich) zurück: Ein Macho ist ein sich betont männlich gebender Mann, der sich Frauen überlegen fühlt bzw. aus der Männlichkeit ausschließlich positive Werte ableitet. Die Ideologie von der Überlegenheit des Männlichen gründet auf der Wertschätzung des Virilen, der männlichen Kraft, Stärke und Potenz bzw. einer Maskulinität (abgeleitet von maskulin; lat.* masculinus*: männlich), die sich durch Aktivität, Tatkraft, Energie usw. auszeichnet. [...]* ■

Mechthilde Vahsen: »Männlich / Männlichkeit / Männlichkeitsforschung«, in: Metzler Lexikon Gender Studies, Geschlechterforschung. Herausgegeben von Renate Kroll. Seite 252-253.

male

*The terms male/masculine refer to something appropriate to man: brave, courageous (*Duden Etymologie *1989, page 438). These reflect a gender ideology in which the male sex is given ›natural‹ attributes. Furthermore, these attributes and sex characteristics – in particular and also in comparison with female and feminine – emphasise power and dominance aspects. [...]*

The every day term of being macho describes a male with excessive feelings of masculinity (cf. Duden Etymologie *1989, page 431); the Mexican word macho derives from Latin* masculus *(male). A macho is a male who acts in an especially masculine fashion, who feels superior to women or who interprets exclusively positive values from masculinity. The ideology of this superiority of the masculine is based on the values associated with virile, male strength, power, potency or masculinity (derived from Latin* masculinus*), that is characterised by activity, vigor, energy etc. [...]* ■

Mechthilde Vahsen. »Männlich / Männlichkeit / Männlichkeitsforschung«, in: Metzler Lexikon Gender Studies, Geschlechterforschung. Edited by Renate Kroll. Page 252-253.

biologischer Determinismus

biologischer Determinismus (lat. determinare*: begrenzen, bestimmen)*

Aus der biologischen (anatomischen, physiognomischen, hormonalen, libidinösen) Konstitution des Menschen kausal abgeleitete Bestimmung seines Wesens und Verhaltens, verknüpft mit der Vorstellung einer leibgebundenen, außersozialen Handlungsdisposition. In wissenschaftlichen Erklärungsansätzen werden insbesondere die unterschiedlichen Körperfunktionen in bezug auf die menschliche Reproduktion zur Erklärung von genuin weiblich bzw. männlich definiertem Verhalten in sozialen Kontaktsituationen herangezogen.

Die Auseinandersetzung mit dem biologischen Determinismus beeinflusste gegen Ende des 20.Jh. die Diskussion dichotomer Begriffe wie sex *und* gender*, Natur und Kultur, Essentialismus und Konstruktivismus. Der biologische Determinismus steht dabei in einem begrifflichen Zusammenhang mit* sex*, mit der Auffassung von Geschlecht als einem ›natürlichen‹, biologisch determinierten, das heißt auch der Geschlechtscharakter bzw. die Geschlechtsidentität wird als essentiell, auf der Basis der biologischen Geschlechtsmerkmale gründend, aufgefasst. (Im Unterschied dazu verweist* gender *auf die Konstruktion des Geschlechts in sozialen und kulturellen Zusammenhängen.)*

Da die biologische Determiniertheit anhand von empirisch-naturwissenschaftlichen Daten erklärt und nicht aus mythisch-religiösen Ideen abgeleitet wird, gehört ihre theoretische Begründung in die Moderne. Seit dem 18.Jh. ▸ *wird*

Biological Determinism

Biological determinism (Lat. determinare: *limiting, determining)*

Biological factors (anatomical, physiognomical, hormonal, libidinousness) in the human constitution causally determine an individual's being and behaviour, linked with the notion of a physical, extra-social behavioural disposition. Scientific explanations refer in particular to different bodily functions with respect to human reproduction to explain genuine defined male or female behaviours in social contact situations.

Towards the end of the 20th century the debate on biological determinism began to influence the discussion of dichotomic terms such as sex *and* gender*, nature and culture, essentialism and constructivism. Biological determinism has a contextual association with* sex*, with the understanding of sex as being ›natural‹, as biologically determined, i.e. that sexual character or gender identity is considered essential, founded on the basis of biological sexual characteristics. (*Gender *differs here by its reference to the construction of sex in social and cultural terms.)*

Because biological determinism roots its explanations on in empirical-scientific data and is not derived from mythical-religious ideas, its theoretical foundations belong to the modern era. Since the 18th century, the description of two gender characteristics [...] have been argued and specified based on the different physical and psychological dispositions of men and women. These are then interpreted as natural gender roles which constitute specific social spheres of ▸

biologischer Determinismus

denn auch die Vorstellung von zwei Geschlechtscharakteren [...] in Form unterschiedlicher physischer und psychischer Dispositionen von Männern und Frauen propagiert und festgeschrieben. Diese werden als natürliche Geschlechterrollen aufgefasst, die einen spezifischen gesellschaftlichen Wirkungsbereich für Männer und Frauen konstituieren, der sozial invariant bleibt.

Der biologische Determinismus wird zum Teil durch eine evolutionsgeschichtliche Interpretation untermauert, nach der in den frühesten Gemeinschaften eine geschlechtsspezifische Arbeitsteilung vorherrschte. Frauen waren durch Gebären und Kinderaufzucht unbeweglicher und beschränkten sich auf häusliche Tätigkeiten, während Männer den Außenraum erkundeten. Diese Ausgangssituation könnte dazu geführt haben, dass sich in der biologisch-körperlichen Evolution die Unterschiede zwischen Frauen und Männern immer weiter ausprägten und sich von daher auch binär angelegte Rollenschemata verfestigten. [...]

Die Geschlechterforschung wiederum lehnt es ab, aus den modernen soziostrukturellen Konstellationen (allgemein und insbesondere das Geschlechterverhältnis betreffend) die Ursituation menschlichen Zusammenlebens abzuleiten und soziale Bezüge von Geschlechtern zu rekonstruieren, deren wahrgenommene Differenz immer schon vorausgesetzt wird [...]

In der Forschung bleibt die Frage aktuell, wie der Stellenwert von Körperlichkeit bzw. Leibgebundenheit als Medium sozialer Erfahrung und soziokultureller Konstruktionen ▸

activity for men and women and are considered socially invariable.

Biological determinism is currently underpinned in part by an evolutionary interpretation, according to which a sexually specific division of labour existed even in the earliest societies. Women were less mobile as a result of childbirth and parenting, limited to domestic activities, whereas men explored the world outside. This initial situation could have resulted in the biological-physical evolutionary differences of men and women becoming more pronounced and more rigidly ingrained in a binary role spectrum. [...]

In contrast, gender research rejects the derivation of primal human communal life from modern socio-structural constellations (in general and in particular that of the relationship between the sexes) and the reconstructing of social bonds between the sexes based on perceived differences long assumed as true. [...]

Within research, the question is still open as to how to interpret the actual significance of physicality as a medium of social experience and socio-cultural constructions. The advancement of technology and rationalisation during the 20th century within the labour process and production of goods and figuratively speaking with respect to social relations resulted in the role of physical activities being played down. New techniques of body forming and reshaping further enable an ongoing abstraction from bodily constitutions previously regarded as natural. The body is, however, considered to be the ultimate guarantee of authenticity; in ►

einzuschätzen ist. Durch die fortschreitende Technisierung und Rationalisierung im 20.Jh. innerhalb des Arbeitsprozesses und der Güterproduktion sowie im übertragenen Sinne in Bezug auf soziale Beziehungen ergab sich ein Bedeutungsverlust körperlicher Aktivitäten. Neue Techniken der Körpergestaltung und -umformung ermöglichen eine weitere Abstraktion von der bisher als natürlich angesehenen Körperkonstitution. Der Körper wird aber auch zum letzten Garanten für Authentizität; zu Beginn des 21.Jh. wird ihm wieder (verstärkt) symbolische Bedeutung im sozialen Kontakt zugeschrieben. Dies wird als Bekräftigung der Theorie des biologische Determinismus gewertet, aber auch als Ausdruck der Wechselbeziehung von kultureller Körperdeutung und Sozialstruktur interpretiert. ■

Kerstin Heyd: »Determinismus, biologischer«, in: Metzler Lexikon Gender Studies, Geschlechterforschung. Herausgegeben von Renate Kroll. Seite 65-67.

the early 21st century it is again (increasingly) attributed with symbolic character within the sphere of social contact. This is interpreted as an argument in favour of the theory of biological determinism but also as an expression of the reciprocity of cultural ›reading of the body‹ and social structures. ■

Kerstin Heyd: »Determinismus, biologischer«, in: Metzler Lexikon Gender Studies, Geschlechterforschung. Edited by Renate Kroll. Page 65-67.

Androgynie (gr. androgynos: *mannweiblich) bezeichnet im medizinischen und biologischen Diskurs ein Individuum ohne eindeutige Geschlechtsausprägung. Die Persönlichkeitspsychologie hat seit den 1970erJahren das Konzept einer psychischen Androgynie entwickelt, mit der Individuen bezeichnet werden, die zusätzlich zu den für ihr eigenes Geschlecht als typisch angesehenen Eigenschaften auch solche des anderen ausgeprägt haben; Androgynie gilt danach als besonders günstige Persönlichkeitsausprägung, da die bipolare Geschlechtertypisierung zu Verhaltenseinschränkungen führe. [...]* ■

Lena Lindhoff: »Androgynie«, in: Metzler Lexikon Gender Studies, Geschlechterforschung. Herausgegeben von Renate Kroll. Seite 9-10.

androgyny / unisex

Androgyny (Greek androgynos: *male-female) describes an individual without clear sexual differentiation in medical and biological literature. Since the 1970s, personality psychology has developed a concept of psychological androgyny to describe individuals who in addition to those characteristics regarded as typical for their sex, also express those of the other. Androgyny is considered as a particularly favourable personality type, since bipolar gender typing results in behavioural restrictions. [...]* ■

Lena Lindhoff: »Androgynie«, in: Metzler Lexikon Gender Studies, Geschlechterforschung. Edited by Renate Kroll. Page 9-10.

Stereotype

Stereotyp / Geschlechterstereotyp (gr. stereos*: starr, unbeweglich; gr.* typos*, lat.* typus*: Muster, Modell)*

Simplifizierte Vorstellung über relevante Eigenschaften einer Personengruppe, die in Verlauf der Sozialisation als kognitive Wissensbestände übernommen werden. Merkmale wie Geschlecht, Rasse und Alter vereinfachen im Dienste einer kognitiven Entlastung die Orientierung und Wahrnehmung der Wirklichkeit und stellen Urteile und Entscheidungen auf der Basis eines raschen Überblicks über begrenzte Informationen bereit. Die Mitgliedschaft in einer sozialen Kategorie führt dazu, dass bestimmte Erwartungen aktiviert werden, die als Stereotype im Sinne von Wahrscheinlichkeitsannahmen wirken und als Rollenerwartungen normativen Charakter tragen können. Geschlechter-Stereotype sind schematische, auf bestimmte Normvorstellungen fixierte Zuschreibungen von Tätigkeiten und Eigenschaften an Frauen und Männer, durch die Verhaltensmöglichkeiten wie z.B. prosoziales /aggressives Handeln je nach Geschlechtszugehörigkeit abgesteckt und Alternativen ausgeblendet werden. ■

Ilka Borchers: »Stereotyp / Geschlechterstereotyp«, in: Metzler Lexikon Gender Studies, Geschlechterforschung. Herausgegeben von Renate Kroll. Seite 377-378.

Stereotype

Stereotype / gender stereotype (Greek stereos*: rigid, immovable; Greek* typos*, Lat.* typus*: pattern, model)*

A simplistic notion of relevant characteristics of a group of persons taken on board as a cognitive knowledge base during the course of socialisation. Characteristics such as sex, race and age help simplify orientation and the perception of reality, help reduce the cognitive burden and are a source of prejudice and decisions made on the basis of a superficial overview of limited information. The membership of a social category results in the activation of certain expectations, which act as a stereotype in the sense of perceptive assumptions and can be normative in character in terms of role expectations. Gender-stereotypes are schematic descriptions of activities and attributes of men and women fixed to specific standard precepts via which behavioural options, such as pro-social / aggressive actions are pegged out depending upon which sexual group one belongs to and via which alternatives are masked. ■

Ilka Borchers: »Stereotyp / Geschlechterstereotyp«, in: Metzler Lexikon Gender Studies, Geschlechterforschung. Edited by Renate Kroll. Page 377-378.

Genderisierung / Doing gender

Der Begriff stammt aus der Ethnomethodologie. Er besagt, dass die ›Herstellung‹ von Geschlecht [...] ein alltäglicher, unvermeidlicher Prozess ist. [...] Ausgegangen wird davon, dass gender in täglichen Interaktionen hergestellt wird und gleichzeitig alltägliche Interaktionen strukturiert. Gender *wird im Kontext einer routinisierten, permanent wiederholten Praxis erworben. Diese Praxis besteht aus Aktivitäten, die sich auf der Ebene der Darstellung sowie der Wahrnehmung als Manifestationen männlicher und weiblicher ›Seinsweisen‹ zeigen. Das Konzept des* doing gender *stützt sich auf die Unterscheidung von* sex *und* gender. *Unter* sex *werden jene (sozial determinierten) biologischen Kriterien verstanden, die Geschlechtszugehörigkeit bestimmen. Sie wird in der alltäglichen Interaktion aber nicht durch den Verweis auf biologische Gegebenheiten hergestellt, sondern durch geschlechtstypische Praktiken.* Doing gender *ist das Einhalten normativer Konzeptionen, die zu einer Geschlechterkategorie ›passen‹. – Ob ein bestimmtes Verhalten zu einer Geschlechterkategorie passt, wird routinemäßig durch den Bezug auf* gender-*Prototypen entschieden. Diese stellen ein Verhaltensrepertoire dar, das regulative Ideale gelungener Männlichkeit und Weiblichkeit anbietet; sie stecken die Grenze ab, die darüber entscheidet, ob ein bestimmtes Verhalten (noch) die Einordnung in ein Geschlecht erlaubt. Der Bezug auf ein ›vergeschlechtlichtes‹ Verhaltensrepertoire ermöglicht es, ein Geschlecht für andere wahrnehmbar darzustellen. Normalerweise verläuft der Prozess der geschlechtlichen Erkennung routinemäßig. Menschen ›sind‹ einfach männlich oder weiblich.* Sex *und* gender *müssen im Prozess des* doing gender *aber nicht immer ›zusammenfallen‹: Die Darstellung des biologischen Geschlechts kann ›verfehlt‹* ▸

Genderisation / Doing gender

This is a term used in ethnomethodology. It states that the ›production /manufacture‹ of gender [...] is a daily and unavoidable process. [...] This assumes that gender is produced by daily interactions and simultaneously structures daily interactions. Gender is acquired in the context of a routinised permanently reiterated practice. This practice comprises activities expressed at the level of presentation and perception as manifestations of male and female modes of being. This concept of doing gender is based on the differentiation of sex and gender. Sex is interpreted as being those (socially determined) biological criteria which determine the gender. It is produced in the interaction, not by reference to biological attributes but by gender-typical practice. Doing gender is the maintenance of normative concepts which ›fit‹ a gender category – whether or not a specific behaviour matches a gender category is decided on a routine basis by reference to gender prototypes. These represent a behaviour repertoire which has a regulative function and presents a number of successful male and female ideals; it demarcates borders which decide whether or not a specific behaviour (still) allows categorisation into a gender. Reference to a ›pre-genderised‹ behaviour repertoire allows a sex to be expressed in a fashion perceivable by others. Normally the process of gender recognition is routine. People – are – simply male or female. Sex and gender do not always have to ›overlap‹ in the process of doing gender. The expression of biological sex can be incorrect; e.g. a woman whose expression of her gender is far removed from one of the prototypes will be considered a male, even if that is not her intention. On the other hand, prototypes do not allow expression of one's ›own‹ but only that of the ›other‹ gender. [...] The orientation ▸

werden; z.B. kann eine Frau, die in ihrer Geschlechtsdarstellung weit von einem Prototyp entfernt ist, für einen Mann gehalten werden, selbst wenn das nicht ihre Absicht ist. Andererseits ermöglichen Prototypen nicht nur die Darstellung des ›eigenen‹ sondern auch jene des ›anderen‹ Geschlechts. [...] Die Orientierung an einem Prototyp von Männlichkeit oder Weiblichkeit kann allerdings auch so überzogen sein, dass eine unzweifelhafte Erscheinung als (›normale/r‹) Frau / Mann nicht bewirkt wird. Die ethnomethodologische Konzeption von doing gender *besteht darauf, dass* gender *omnipräsent und omnirelevant ist, denn die Gesellschaft beruht auf der Vorstellung der Zweigeschlechtlichkeit und darauf, dass die Geschlechter unterschiedliche Einstellungen, Verhaltensweisen und Eigenschaften haben. Das Verhalten eines Menschen wird immer auch als das eines Mannes oder einer Frau wahrgenommen und eingeschätzt. Das biologische und das soziale Geschlecht werden in der alltäglichen Praxis zugeschrieben; Verhalten wird danach bewertet, ob es der Darstellung eines Geschlechts angemessen ist oder nicht. –* Gender *wird als dynamische Kategorie angesehen;* doing gender *bedeutet nicht notwendigerweise, dass normative Konzeptionen von Männlichkeit und Weiblichkeit erfüllt werden müssen. Überschreitungen von Konventionen können neue Konventionen begründen. So wird der Sinn von* gender *differenziert, die Möglichkeiten der willkürlichen und der unwillkürlichen Darstellung werden erweitert.* ■

Elisabeth Holzleithner: »Genderisierung / Doing Gender«, in: Metzler Lexikon Gender Studies, Geschlechterforschung. Herausgegeben von Renate Kroll. Seite 72-73.

to a prototype matching the male or female image can, however, become extreme to the extent that an undisputed appearance as a ›normal› man /woman is not achieved. The ethnomethodological concept of doing gender *states that* gender *is omnipresent and omnirelevant, because society is based on the notion of dual genders and genders have different attitudes, behaviours and characteristics. A person's behaviour is perceived and interpreted as being that of either a man or a woman. The biological and the social gender are attributed on a daily basis: behaviour is therefore judged as to whether or not it is appropriate to one of the genders or not.* Gender *is viewed as a dynamic category;* doing gender *does not necessarily imply that the normative concepts of masculinity and femininity have to be met with. Transgression of conventions can result in new conventions. The sense of* gender *becomes more differentiated; the opportunities for arbitrary and spontaneous representation are extended.* ■

Elisabeth Holzleithner: »Genderisierung / Doing Gender«, in: Metzler Lexikon Gender Studies, Geschlechterforschung. Edited by Renate Kroll. Page 72-73.

Geschlechterrollen

Der Begriff stammt aus der soziologischen Rollentheorie. Er bezeichnet die Summe der von einem Individuum erwarteten Verhaltensweisen als Frau bzw. als Mann und damit ein überindividuelles, relativ stabiles und insofern vorhersagbares geschlechtsspezifisches Verhaltensmuster. [...] Geschlechterrollen werden im Prozess der primären Sozialisation erlernt und so internalisiert bzw. persönlichkeitsstrukturell verankert, dass sie zur Ausbildung einer Geschlechtsidentität führen. [...]

Eine Diskriminierung aufgrund von Geschlechterrollen manifestiert sich insbesondere in starren Weiblichkeitsbildern bzw. -stereotypen, die als defizitäres Gegenbild zu einer positiven männlichen Norm konstruiert und im soziokulturellen System immer wieder reproduziert werden. [...]

Die seit dem 18.Jh. in der bürgerlichen Gesellschaft vorherrschende Vorstellung von binären Geschlechterrollen und – damit verbunden – von stabilen männlichen und weiblichen ›Geschlechtscharakteren‹ basiert auf der Trennung zwischen Privatheit und Öffentlichkeit.

Demgegenüber werden Geschlechterrollen in der Moderne zurückgewiesen oder umgedeutet und feste Geschlechterpositionen in Frage gestellt. Die Postmoderne geht von multiplen und dynamischen gendered identities im Sinne von kulturellen Einschreibungen aus, mit denen Individuen ein imaginäres Verhältnis zu ihren Lebensbedingungen konstruieren. [...] ■

Doris Feldmann / Ina Habermann: »Geschlechterrolle«, in: Metzler Lexikon Gender Studies, Geschlechterforschung. Herausgegeben von Renate Kroll. Seite 158-159.

Gender Roles

This is a term from sociological role theory. It describes the sum of expected behaviours for an individual as a woman or a man and is hence a super-individual, relatively stable and thus predictable gender-specific behaviour pattern. [...] Gender roles are learnt in a process of primary socialisation, becoming internalised or anchored within a personality structure such that they lead to the formation of a gender identity. [...]

Discrimination as a consequence of gender roles is manifested in particular in rigid feminine images or stereotypes constructed as a deficient counterpart to a positive male standard and which is repeatedly reproduced within socio-cultural systems. [...]

The prevailing notion of binary gender roles as has been predominant since the 18th century in bourgeois society – and hence related to stable male and female ›gender characters‹ founded on the distinction between private and public.

In contrast, gender roles are rejected or reinterpreted in the modern age and rigid gender positions put into doubt. In the post-modern era, the assumption is of multiple and dynamic gender identities in the sense of cultural inscriptions with which individuals create an imaginary relationship to their lifestyle. [...] ■

Doris Feldmann / Ina Habermann: »Geschlechterrolle«, in: Metzler Lexikon Gender Studies, Geschlechterforschung. Edited by Renate Kroll. Page 158-159.

Geschlechterdifferenz

Geschlechterdifferenz bezeichnet die Verschiedenheit der Geschlechter bzw. Geschlechterrollen; der Begriff steht im Zusammenhang mit der Definition von Geschlecht und Geschlechtsidentität bzw. mit einer Geschlechterordnung, die nicht von der Gleichheit der Geschlechter ausgeht.

Biologisch bezieht sich Geschlechterdifferenz auf die unterschiedlichen anatomischen, physiognomischen, hormonalen, libidinösen Merkmale des menschlichen Körpers (biologischer Determinismus) bzw. auf die unterschiedliche Beteiligung der Geschlechter an der Reproduktion bzw. Reproduktivität. Geschlechterdifferenz lässt sich aber auch philosophisch, historisch, politisch, sozial, kulturell, morphologisch, sprachlich und diskursiv begründen. Im feministischen Diskurs verweist der Begriff Geschlechterdifferenz auf das Phänomen der Männerzentriertheit (Androzentrik) auf allen Ebenen des gesellschaftlichen und kulturellen Lebens, auf eine über Räume und Zeiten hinweg bestehende Hierarchie der Geschlechter. Das heißt auch, der an sich wertneutrale Begriff Geschlechterdifferenz impliziert geschlechtsspezifische Asymmetrien, Marginalisierung, Unterdrückung. Das zeitgenössische Verständnis von Geschlechterdifferenz ist ein Produkt der Moderne. [...]

Mit der empirisch-sensualistischen Definition der Geschlechter in der Aufklärung, mit der Industrialisierung und dem Wandel sozio-kultureller Strukturen wurde (auch) die biologische Begründung der polarisierten Weiblichkeits- und Männlichkeitsvorstellungen nötig: die Zweigeschlechtlichkeit wurde propagiert. Es kam zu einer (für die kulturelle Moderne konstitutiven) Codierung der Geschlechter, d.h. ▶

Gender Difference

Gender difference denotes the difference between sexes and gender roles; the concept is related to the definition of sex and gender identity or with that of a gender order which does not assume equality of the sexes.

In biological terms, gender difference assumes different anatomical, physiognomical hormonal and libidinousness attributes of the human body (biological determinism) or according to the different roles played by the sexes in reproduction and reproductivity. Gender difference can also be founded philosophically, historically, politically, socially, culturally, morphologically, linguistically and discursively. Within the feminist discourse, reference is made to the term gender difference as a phenomenon of male centricity (androcentrism) at all levels of social and cultural life, to an existing hierarchy of the sexes over time and space. I.e., also the intrinsically neutral term gender actually implies gender-specific asymmetries, marginalisation, repression. The contemporary understanding of gender difference is a product of modernism. [...]

The empirical-sensual definition of the sexes in education, together with industrialisation and the transition of socio-cultural structures, went hand-in-hand with the biological arguments and polarised feminine and masculine notions became necessary: propagating the concept of dual genders. This gave rise to a (constitutive for cultural modernism) coding of the genders, i.e. establishing gender differences. It was based on contrasts between nature (as noted by Aristotle) (nature in terms of materials, feelings, passions) and culture (form, understanding, judgement), allocated in gender- ▶

Geschlechterdifferenz

zu einer Festschreibung der Geschlechterdifferenz. Sie basiert auf dem (bereits von Aristoteles konstatierten) Gegensatz von Natur (Materie, Gefühl, Leidenschaft) und Kultur (Form, Verstand, Urteil), der geschlechtsspezifisch zugeordnet wurde – ein Konzept, das mit tradierten Vorstellungen bzw. Vorurteilen vom sogenannten ›schwachen‹ bzw. ›starken‹ Geschlecht übereinstimmte. In der älteren wie in der Neuen Frauenbewegung gibt es Strömungen, in denen die bewusste Verwendung der Geschlechterdifferenz als ›Sprechposition‹ von Frauen auf ihre gemeinsame Identität als Genusgruppe bzw. ihre Erfahrungen mit der männlichen Herrschaft verweisen soll. Das Geschlechterverhältnis wird in diesen Zusammenhängen als das älteste und fundamentalste Herrschaftsverhältnis angesehen. Vertreterinnen von Theorie und Politik der Geschlechterdifferenz wenden sich gegen traditionelle Rollenaufteilungen und androzentrische Definitionen von Geschlechterdifferenz bzw. von Weiblichkeit; sie plädieren für die Selbstbestimmung der Geschlechter und für die Anerkennung der Geschlechterdifferenz als ›demokratische Differenz‹. [...] ■

Renate Kroll: »Geschlechterdifferenz«, in: Metzler Lexikon Gender Studies, Geschlechterforschung. Herausgegeben von Renate Kroll. Seite 153-154.

specific terms – a concept which fits in with the traditional notions and prejudices of the so-called ›weak‹ and ›strong‹ sexes. In both older as well as more recent women's movements, there are streams in which the differences between the sexes are consciously used by women to ›take a stand‹, to spotlight their shared identity of a genus group or their experiences under male domination. The relationship between the sexes in this sense is considered to be the oldest and most fundamental relationship of dominance. Representatives of the theory and politics of gender difference have turned against traditional role distributions and androcentric definitions of gender difference or of femininity; they argue for self-determination of the genders and for recognition of the gender difference as a ›democratic‹ difference. [...] ■

Renate Kroll: »Geschlechterdifferenz«, in: Metzler Lexikon Gender Studies, Geschlechterforschung. Edited by Renate Kroll. Page 153-154.

Gender Mainstreaming

Neuer Ansatz in der Geschlechtergleichstellungspolitik: Die gender-*Frage soll in alle Bereiche des politischen Lebens hineingetragen und nicht nur sektoral, in Fraulabteilungen oder durch Gleichstellungskomissionen, behandelt werden. So kommen ›neue‹ Akteur/innen ins Spiel: Entscheidungsträger/innen, die an den Schalthebeln der Macht sitzen. Sie sollen bei allen politischen Maßnahmen in Form einer Art »Geschlechterverträglicheitsprüfung« (*Gender Impact Assessment, GIA*), die* gender-*Frage berücksichtigen. Der Begriff tauchte zum ersten Mal bei der 4. Weltfrauenkonferenz (1985) in Nairobi im Kontext der Entwicklungszusammenarbeit auf. Als übergreifendes Konzept wurde* gender mainstreaming *auf der 5. Weltfrauenkonferenz (1995) in Beijing postuliert. Der Europarat und die Europäische Union haben dieses Postulat als zentrales Element in ihre Politik übernommen, siehe Art. 3 des EG-Vertrags, nach dem die Europäische Gemeinschaft bei allen Maßnahmen auf die Beseitigung von Ungleichheiten und auf die Förderung der Gleichstellung von Frauen und Männern hinzuwirken hat. Gender mainstreaming soll die konventionellen Ansätze der Geschlechtergleichstellungspolitik nicht ersetzen, sondern sie ergänzen. [...]* ■

Elisabeth Holzleithner: »Gender mainstreaming«, in: Metzler Lexikon Gender Studies, Geschlechterforschung. Herausgegeben von Renate Kroll. Seite 142-143.

Gender Mainstreaming

This is a new approach to the policies of gender equality: The gender *issue is to be opened up and introduced into all areas of political life, not only sectorally in women's departments or in equality committees. This concept brings ›new‹ players into the field: decision makers (male and female) who sit at the levers of power. They are to consider the* gender *issue in all political actions by way of a* gender impact assessment, GIA. *The term* gender mainstreaming *first arose during the 3rd World Conference on Women (1985) in Nairobi in the context of development co-operation.* Gender mainstreaming *was then postulated as an overarching concept at the 4th World Conference on Women (1995) in Beijing. The European Council and the European Union have accepted this postulate as a central element of their policies, see also article 3 of the EC treaty, according to which the European Community has to seek in all of its activities to remove inequalities and promote the equality of men and women. Gender mainstreaming is not intended as a replacement for conventional approaches to gender sex equalisation policies, but as a complement.* ■

Elisabeth Holzleithner: »Gender mainstreaming«, in: Metzler Lexikon Gender Studies, Geschlechterforschung. Edited by Renate Kroll. Page 142-143.

Gender Design

[...] Gender Design widmet sich der Analyse der Objekte (Objekt als Oberbegriff für alle gestalteten Produkte, Zeichen, Konzepte, Prozesse) sowie der Beziehung zwischen Subjekt und Objekt unter der Perspektive ihrer Vergeschlechtlichung. Professionelle Designer gestalten – bewusst oder unbewusst – als sozial und kulturell geprägte Geschlechtswesen, gleichzeitig kommen in der Nutzung von Design alltäglich vergeschlechtlichte Handlungs- und Verhaltensweisen zum Ausdruck, drücken sich Projektionen und Wünsche immer auch geschlechtlich konnotiert aus. Solche zum Teil verdeckten oder negierten Phänomene des Gendering werden unter dieser Perspektive systematisch aufgezeigt, und ihre Auswirkungen auf das Design werden ergründet, um sie zu einem selbstverständlichen Bestandteil jeglicher Design-Forschung und -Praxis zu machen. [...] ■

Uta Brandes in: Wörterbuch Design, Begriffliche Perspektiven des Design
Michael Erlhoff, Tim Marshall (Hg.)
Board of International Research in Design, BIRD
2008 Birkhäuser Verlag AG Basel, Seite 172-173

Gender Design

[...] Gender Design focus on the analysis of objects (whereas object is a generic term referring to designed products, signs, concepts, and processes in their entirety), and relations between subject and object under the perspective of gendering. Professional designers create and design – conscious or unconscious – as socially and culturally characterized by gender; contemporary gendered behavior is given expression in everyday usage of design and projections and wishes always manifest gender connoted. Such, for some part hidden or negated, phenomena of gendering are systematically shown under this perspective, and their influence on design is comprehended to become a natural part of research and practice of Design. [...] ■

Uta Brandes in: Wörterbuch Design, Begriffliche Perspektiven des Design
Michael Erlhoff, Tim Marshall (ed.)
Board of International Research in Design, BIRD
2008 Birkhäuser Verlag AG Basel, p. 172-173

Gender Codes

Das Ausblenden von Gender ist eine große Herausforderung. Da kulturell kein Zeichenvorrat für Androgynität verankert ist, kann eine geschlechtsneutrale Anmutung nur dann erreicht werden, wenn »weibliche« und »männliche« Codes zu einer Einheit verschmelzen oder wenn sie ausgewogen nebeneinander auftreten. ■

aus: Haslinger, Susanne
Gender & Design – Leitfragen
Hrsg. Zentrum Frau in Beruf und Technik (ZFBT), Castrop-Rauxel 2006

Gender Codes

It is the blending out of gender that is the huge challenge. Since androgyny has no anchored cultural images, a gender-neutral appearance can only be achieved if the codes for »male« and »female« are blended into a single entity or when they occur in balance, next to one another. ■

from: Haslinger, Susanne
Gender & Design – Leitfragen (Key questions)
Published by Zentrum Frau in Beruf und Technik (ZFBT), Castrop-Rauxel 2006

Genderspezifisches Design

[...] Uns hat die Frage interessiert, welcher Mensch sich vor dem geistigen Auge manifestiert, wenn es um Produkte, Dienstleistungen und Ansprachekonzepte geht: Ist es jener androgyne Mensch, der in all seinen Facetten – ob jung, ob alt, ob arm, ob reich, ob groß, ob klein, ob laut, ob leise, ob Mann, ob Frau – akzeptiert wird? Unser Urteil lautete: Ja, manchmal – aber viel zu selten. Wir konnten oft, viel zu oft, stereotype Bezugnahmen auf Männer und Frauen und darauf, wie sie sein sollten, feststellen. ■

aus: Haslinger, Susanne
Gender & Design – Leitfragen
Aus dem Vorwort von Petra Kerstin und Bettina Vaupel
Hrsg. Zentrum Frau in Beruf und Technik (ZFBT), Castrop-Rauxel 2006

Gender-specific Design

[...] We were interested in the question of which person appears in one's mind's eye when thinking of products, services and marketing concepts: is it some androgynous being accepted in all of its facets – young or old, poor or rich, large or small, loud or quiet, man or woman?
We think the answer is yes, sometimes – but all too seldom. We were able, all too frequently, to find stereotyped references to men and women and how they should ideally be. ■

from: Haslinger, Susanne
Gender & Design – Leitfragen (Key questions)
Extract from the foreword by Petra Kerstin and Bettina Vaupel
Published by Zentrum Frau in Beruf und Technik (ZFBT), Castrop-Rauxel 2006

Gendersensibles Design

Die Basis ist ein demokratisches Designverständnis und die Frage nach der Notwendigkeit von geschlechterdifferenzierter Gestaltung von Produkten und Prozessen. Das Ziel ist eine Gestaltung, die Chancengleichheit der Geschlechter ermöglicht und jede Aus- oder Abgrenzung vermeidet. Differenzierungen sind entsprechend der Bedürfnisvielfalt des Menschen, der Funktion und der unterschiedlichen ergonomischen Erfordernisse zu entwickeln. Dazu ist die stereotype Gestaltung zu vermeiden und Gender Codes nur reflektiert einzusetzen. Gendersensibles Design erfordert die Genderkompetenz von Entwicklern und Designern. ■

Birgit Weller und Katharina Krämer

Gender-sensitive Design

The basis is built on a democratic comprehension of design, and a question of the necessity on gender differentiating design of products and processes. The aim is a design that offers equality of opportunities without any distinction or exclusion. Differentiations are made accordingly to the diversity of people, function and ergonomics. A stereotypical design should be avoided and gender codes should only be used in a reflected manner. Gender sensitive design does require gender competence of developers and designers. ■

Birgit Weller and Katharina Krämer

Diversity

Gleiche Chancen für alle – unabhängig von Geschlecht und Hautfarbe. Dieser Anspruch gilt [...] weltweit und ist in unseren Grundwerten und Leitlinien fest verankert. Kein Arbeitnehmer sollte wegen seiner Hautfarbe gemieden oder vorgezogen werden, keine Mitarbeiterin weniger Lohn erhalten als ein männlicher Kollege in der gleichen Position. ■

Definitionen des Bundesministeriums für Familie, Senioren, Frauen und Jugend
www.bmfsfj.de

Diversität ist ein Zusage an die Wahrnehmung und Wertschätzung der Vielzahl an Charakteristika, die Individuen einzigartig machen in einer Atmosphäre, in der individuelle und kollektive Errungenschaften gefördert und gefeiert werden.

Beispiele dieser Charakteristiken sind: Alter; Wahrnehmung; Kultur; Benachteiligungen, ob mental, physisch oder das Lernverhalten betreffend; sozio-ökonomische Herkunft; gesprochene Sprache(n); Familienstand; Aussehen; politische Gesinnung; Herkunft; Glaube oder sexuelle Orientierung. ■

The University of Tennessee Libraries: Diversity Committee, Frühjahr 2011; überarbeitet Januar 2003
http://www.lib.utk.edu/diversity/diversity_defintion.html

Diversity

Equal opportunities for all – not matter what gender or color of skin. This claim is valid worldwide and is firmly fixed in our principles and policy guidelines. Employees must not be dismissed or give undue preference to due to their color of skin, a female employee must not draw less salary than an male colleague. ■

Definition by the Federal Ministry for Families, Senior Citizens, Women and Youth
www.bmfsfj.de

Diversity is a commitment to recognizing and appreciating the variety of characteristics that make individuals unique in an atmosphere that promotes and celebrates individual and collective achievement.

Examples of these characteristics are: age; cognitive style; culture, disability (mental, learning, physical); economic background; education; ethnicity; gender identity; geographic background; language(s) spoken; marital/partnered status; physical appearance; political affiliation; race; religious beliefs; sexual orientation. ■

The University of Tennessee Libraries Diversity Committee, Spring 2001; Revised January 2003
http://www.lib.utk.edu/diversity/diversity_definition.html

Literaturempfehlungen
Further Readings

Literaturempfehlugen | Further Readings

de Beauvoir, Simone
Das andere Geschlecht. Sitte und Sexus der Frau
1949, Neuausgabe 2000, Rowohlt Verlag Reinbek
ISBN 978-3-499-22785-1

Bieling, Tom
Gender Puppets - Geschlechterinszenierung anhand der nonverbalen Kommunikation von Schaufensterpuppen.
Kölner Internationale Schriften zum Design (Band 1)
LIT Verlag 2008
ISBN-10: 3825812456
ISBN-13: 978-3825812454

Brandes, Uta
Die Geschlechtersprache der Produkte
in: Zeitschrift für Frauenforschung und Geschlechterstudien
20. Jahrgang, Heft 4/2002, Kleine Verlag Bielefeld
ISSN 0946-5596

Brandes, Uta
Dazwischen. Design und Geschlecht.
in: Das Undisziplinierte Geschlecht
Opladen, Leske+Budrich 2000

Brandes, Uta
Designing Gender: Das Drama der Geschlechter in Logo-Gestaltungen
in: Werbung, Mode und Design
Wiesbaden, Westdeutscher Verlag 2001

Brandes, Uta
androgyn – unisex – geschlechtslos: unscharfe Geschlechter.
in: Körpernah – körperfern – Modellierung des Weiblichen in der Mode
Museum für Angewandte Kunst, Köln 2003

Brandes, Uta; Stich, Sonja
Über die unbewusste und bewusste Vergeschlechtlichung von Produkten.
Eine (meta-) morphologische Studie in zwei Teilen
Köln 2001/2003, unveröffentlicht

von Braun, Christina; Stephan, Inge
Gender-Studien
Metzler Verlag Stuttgart 2000
ISBN 3-476-02143-2
ISBN 978-3-476-02143-4

Buchmüller, Sandra; Joost, Gesche
Erst war das Wort, dann die Form
in: Erlhoff, Michael (Hrsg): Uta Brandes – Sie, und sie als Novelle
Walter König, Köln 2009
ISBN 978-3-86560-581-8

Buchmüller, Sandra
Gendered Interfaces. Zum Einfluss digitaler Benutzungsoberflächen auf virtuelle Geschlechterbilder in Online-Rollenspielen
in: JFC Medienzentrum Köln (Hrsg.): MedienConcret, Magazin für die Medienpädagogische Praxis: Mediengeneration. Jungen und Mädchen in der Jugend und Mädchenkultur
Köln, Juni 2008

Buchmüller, Sandra
Gendered by Design. Zur kulturellen Konstruktion von Geschlecht durch Design
in: Gender & Design – Hochschule für angewandte Wissenschaft und Kunst (HAWK)
Maria-Goeppert-Mayer Programm, Hildesheim 2008

Doderer, Yvonne P.
Doing Beyond Gender. Interviews zu Positionen und Praxen in Kunst, Kultur und Medien
Monsenstein & Vannerdat, Münster 2008
ISBN 978-3-86582-633-6

Literaturempfehlugen | Further Readings

Doetsch, Brigitte; Henze, Regina; Lesemann, Silke (Hg.)
Technik im Korsett – männlich, weiblich, menschlich!?
2008 Kleine Verlag GmbH, Bielefeld
ISBN 978-3-89370-444-6

Erlhoff, Michael; Marshall, Tim (Hg.), Board of International Research in Design, BIRD
Wörterbuch Design, Begriffliche Perspektiven des Design
2008 Birkhäuser Verlag AG Basel
ISBN 978-3-7643-7739-7

Haslinger, Susanne
Gender & Design – Leitfragen
Hrsg. Zentrum Frau in Beruf und Technik (ZFBT), Castrop-Rauxel 2006

Himmel & Jord GmbH (Hg.)
Rosa für Mädchen und Blau für Jungs – Design aus der Genderperspektive betrachtet
In: Nord No.4, Nord Magazin 2007
Himmel & Jord GmbH Berlin 2007

Jahnke, Marcus (Redaktion)
FormgivningNormgivning
CFK Center for Consumer Science
Handelshochschule der Universität Göteborg, Schweden 2006
ISBN 91-974642-4-4

Joost, G. (Hrsg.)
Designing Gender
in: Gender & Design Hochschule für angewandte Wissenschaft und Kunst (HAWK)
Maria-Goeppert-Mayer Programm
Hildesheim 2008

Joost, Gesche
Do it Yourself. Über die Beziehung von Gestaltung und Geschlecht
in: VIER 08, Magazin der Hochschule für Künste Bremen 2008

Joost, Gesche
Gender-Aspekte in der Designausbildung
HAWK Hildesheim Ringvorlesung 2007/2008
ISBN 978-3-8258-1402-1

Joost, G., Bessing, N., Buchmüller, S.
G – Gender Inspired Technology
in: Ernst, W. (Hrsg.): Geschlecht und Innovation. Gender Mainstreaming im Techno-Wissenschaftsbetrieb
Internationale Frauen- und Genderforschung in Niedersachsen, Teilband 4
LIT Verlag Berlin 2010
ISBN 978-3-643-10712-1

Kirkham, Pat (Hrsg.)
The Gendered Object
Manchester University Press
ISBN 0719044758

Kroll, Renate (Hg.)
Metzler Lexikon Gender Studies, Geschlechterforschung, Ansätze – Personen – Grundbegriffe
2002 J.B. Metzlersche Verlagsbuchhandlung und Carl Ernst Poeschel Verlag GmbH in Stuttgart
ISBN 3-476-01817-2

Lorber, Judith
Gender – Paradoxien
Leske + Budrich, Opladen 1999
ISBN 3-8100-3743-5

Moss, Gloria
Gender, Design and Marketing: How Gender Drives Our Perception of Design and Marketing
Gower Publishing Co Ltd 2009
ISBN-10 0566087863
ISBN-13 978-0566087868

Literaturempfehlugen | Further Readings

Schade, Sigrid; Strunk, Marion (Hg.)
Unterschiede. Unterscheiden. Zwischen Gender und Kulturen
mit dem Beitrag Entscheidende Unterschiede. Asiatische Geschlechterkonstruktionen von Mobiltelefonen von Uta Brandes, Hochschule für Gestaltung und Kunst Zürich 2003
ISBN 3-906437-10-8

Süddeutsche Zeitung Magazin
Süddeutsche Zeitung Magazin(in) – Das Frauenheft
Magazin Verlagsgesellschaft Süddeutsche Zeitung mbH 18.04.2008

http://www.design-research-lab.org/

http://www.doingbeyondgender.net/cms/de/

http://www.femaleinteraction.com/

http://www.femmeden.com/

http://www.gender-design.com/

http://www.gender-happening.de/

http://www.gender-inspired-technology.org/

http://www.genderdesign.info/

http://www.genderstudies-hamburg.de/

http://www.inspiration-verpackung.de/gender-design/

http://www.kisd.de/subject_gd.html

http://www.thing-hamburg.de/index.php?id=933

http://www.uni-due.de/genderportal/

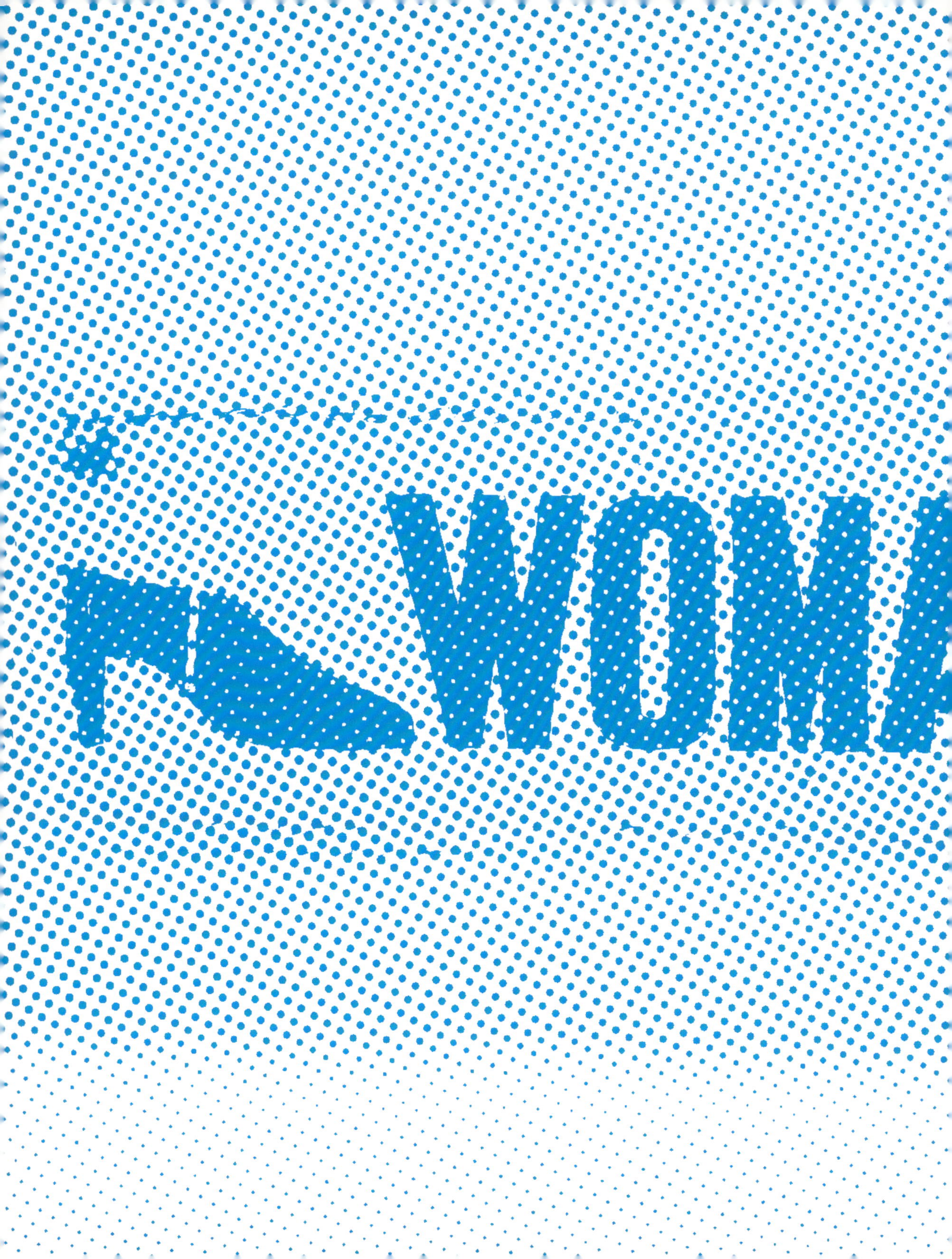
WOM

Index

252

01 Petty & Pink Collection; Esprit
Urban Street Collection; Esprit

Notizbücher DIN A5 | Notebooks DIN A5

Esprit Europe GmbH
Esprit-Allee
40882 Ratingen
Germany
www.esprit.com

Bild | Picture
01, 02, 03, 04: Bildrechte Fotografien | Photography rights Weller, Hinz, Krämer, Rajabi

02 Glamour & Style Collection; Esprit
Blue & Blue Collection; Esprit

Spardosen | Money boxes

Esprit Europe GmbH
Esprit-Allee
40882 Ratingen
Germany
www.esprit.com

Bild | Picture
05, 06, 07: Bildrechte Fotografien | Photography rights Weller, Hinz, Krämer, Rajabi

03 Fuel for Life Woman; Diesel
Fuel for Life Man; Diesel

Duft | Fragrance

Diesel S.p.A.

via dell'Industria, 4/6 n7
36042 Breganze (VI)
Italy
diesel-fragrance-factory.com

Fußnote | Footnote
03.1, 03.2: www.diesel-parfums.com/en/#!/fuel-for-life/unlimited

Bild | Picture
08, 09, 10, 11, 12, 13, 14: Bildrechte Fotografien | Photography rights Weller, Hinz, Krämer, Rajabi

04 Coca-Cola light Coca-Cola zero

Erfrischungsgetränke | Soft drinks

Coca-Cola GmbH
Quartier 205
Friedrichstraße 68
10117 Berlin
Germany
www.coke.de

Fußnote | Footnote
04.1: Pressemitteilung Berlin, Juli 2008: Coca-Cola light setzt mit neuer Kampagne »nur du bist du« auf Individualität, www.coca-cola-gmbh.de/presse/pressemitteilungen/mitteilung/pressrelease.do?id=30599
04.2: Pressemitteilung Berlin, 18. Februar 2008: Coca-Cola Zero startet das Coke Zero Game zur UEFA EURO 2008TM, www.coca-cola-gmbh.de/presse/pressemitteilungen/mitteilung/pressrelease.do?id=26594
04,3: Pressemitteilung Berlin, Juli 2008: Nationale Handels-Promotion zusammen mit Coke Zero, www.coca-cola-gmbh.de/presse/pressemitteilungen/mitteilung/pressrelease.do?id=30599

Bild | Picture
15, 16: Bildrechte Fotografien | Photography rights Weller, Hinz, Krämer, Rajabi

05 Camee Toiletries bag lips; Vlieger & Vandam
Camee Toiletries bag moustache; Vlieger & Vandam

Kulturtaschen | Toiletries bags

Vlieger & Vandam
Nieuwe Herengracht 141 hs.

1011 SG Amsterdam
The Netherlands
www.vliegervandam.com

Bild | Picture
17, 18: Bildrechte Fotografien | Photography rights Weller, Hinz, Krämer, Rajabi

06 EVE; Eve Getränke-GmbH
Holsten Pilsener; Holsten Brauerei

Apéritif Spezialität | Aperitif speciality
Pilsener Bier | Pilsener beer

Eve von Cardinal
www.cardinal-eve.ch

Feldschlösschen Getränke AG
Theophil-Roniger-Strasse
4310 Rheinfelden
Switzerland
www.feldschloesschen.com

Carlsberg Deutschland GmbH
Holstenstraße 224
22765 Hamburg
Germany
www.carlsbergdeutschland.de

Holsten-Brauerei AG
Holstenstraße 224

22765 Hamburg
Germany
www.holsten-pilsener.de

Fußnote | Footnote
06.1: Medienmitteilung Freiburg/Rheinfelden, September 2006: Das neue »EVE«: Cardinal lanciert einen Aperitif für Frauen www.cardinal-eve.ch/medien/bilder/Medienmitteilung_EVE.pdf
06.2: Medienmitteilung Freiburg / Rheinfelden, 02. Juni 2008: Zur neuen Kampagne mit Mahara Mc Kay eine halbe Million Fläschchen EVE von Cardinal www.cardinal-eve.ch/medien/bilder/2008/Medienmitteilung_%20EVE_Neue_Kampagne.pdf
06.3: www.holsten-pilsener.de/index01.php

Bild | Picture
19, 20, 22, 23: Bildrechte Fotografien | Photography rights Weller, Hinz, Krämer, Rajabi
21: Werbedarstellung Eve Getränke-GmbH

07 Anti Aging Bier; Klosterbrauerei Neuzelle
Marathon; Klosterbrauerei Neuzelle

Bier | Beer

Klosterbrauerei Neuzelle GmbH
Brauhausplatz 1
15898 Neuzelle
Germany
www.klosterbrauerei.com/start.html

Fußnote | Footnote
07.1: www.klosterbrauerei.com/html.php/modul/HTMLPages/pid/142
07.2: Produktbeschreibung Etikett | Product description label

Bild | Picture
24, 25, 26, 27: Bildrechte Fotografien | Photography rights Weller, Hinz, Krämer, Rajabi

08 B.United Jeans Woman; Benetton
B.United Jeans Man; Benetton

Duft | Fragrance

Benetton Group Spa
Villa Minelli

31050 Ponzano, Treviso
Italy
www.benetton.com

Fußnote | Footnote
08.1: www.nstperfume.com/2007/09/09/benetton-bunited-jeans-hard-candy-fragrance-new-perfumes

Bild | Picture
28, 29: Bildrechte Fotografien | Photography rights Weller, Hinz, Krämer, Rajabi

09 Pearl & Beauty, Double Effect; Nivea
Dry Impact, Cool Kick; Nivea

Deodorant

Beiersdorf AG
Unnastraße 48
20245 Hamburg
Germany
www.beiersdorf.de

Bild | Picture
30, 31, 32: Bildrechte Fotografien | Photography rights Weller, Hinz, Krämer, Rajabi

10 Vital f; Orthomol
Vital m; Orthomol

Nahrungsergänzungsmittel | Food supplement

Orthomol pharmazeutische Vertriebs GmbH
Herzogstraße 30
40764 Langenfeld
Germany
www.orthomol.de

Fußnote | Footnote
10.1: www.orthomol.de/desktopdefault.aspx/tabid-52/174_read-134
10.2: www.orthomol.de/desktopdefault.aspx/tabid-54/175_read-135

Bild | Picture
33, 34: Bildrechte Fotografien | Photography rights Weller, Hinz, Krämer, Rajabi

11 Venus Breeze; Gillette
Mach 3; Gillette

Nassrasierer | Wet razor

Procter & Gamble Germany GmbH & Co Operations oHG
Sulzbacher Str. 40
65824 Schwalbach am Taunus
Germany
www.pg.com
www.gillettevenus.de
www.gillette.com

Fußnote | Footnote
11.1: www.gillettevenus.de/de_DE/products/breeze/index.jsp
11.2: www.gillettevenus.de/de_DE/tips_and_faqs/moms_needtoknow/index.jsp
11.3: www.gillette.com/de/de/Products/Razors/mach3-turbo/mach3-turbo.aspx
11.4: www.gillette.de

Bild | Picture
35, 36, 39, 40: Bildrechte Fotografien | Photography rights Weller, Hinz, Krämer, Rajabi
37, 38: Werbedarstellungen Procter & Gamble Germany GmbH & Co Operations oHG

12 Supreme, Go Fresh, Beauty; Dove
Extra Fresh, Clean Comfort, Aqua Impact; Dove Men+Care

Duschgel | Shower gel

Unilever Deutschland GmbH
Strandkai 1
20457 Hamburg
Germany
www.unilever.com
www.dove.de
www.dovemencare.de

Fußnote | Footnote
12.1: www.dovemencare.de/de/de/showergel/clean_comfort_pflegedusche/product.html

Bild | Picture
41, 42, 43, 44: Bildrechte Fotografien | Photography rights Weller, Hinz, Krämer, Rajabi

13 Contradiction; Calvin Klein
Contradiction for men; Calvin Klein

Duft | Fragrance

Calvin Klein inc.
205 West 39th Street
New York, 10018
USA
www.calvinklein.com

Fußnote | Footnote
13.1: www.douglas.de/douglas/Calvin-Klein/Contradiction/index_b003210.html
13.2: www.douglas.de/douglas/Calvin-Klein/Contradiction-Men/index_b003211.html

Bild | Picture
45, 46, 47: Bildrechte Fotografien | Photography rights Weller, Hinz, Krämer, Rajabi

14 Telephone Plush pink; Silly Gifts, Present Time
T3, Jacob Jensen

Telefone | Telephones

Present Time bv
Veluwezoom 32
1327 AH Almere
The Netherlands
www.presenttime.com

Jacob Jensen Brand ApS
Hejlskovvej 104
7840 Højslev
Danmark
www.jacobjensen.com

Fußnote | Footnote
14.1: www.dinamo.de/rosa-pluеschtelefon.html
14.2: www.amazon.de/Jacob-Jensen-Telephone-schnurgebundenes-Telefon/dp/B000OW7ZXE

Bild | Picture
48, 49: Bildrechte Fotografien | Photography rights Weller, Hinz, Krämer, Rajabi

15 LEGO® Bauspaß für Mädchen
LEGO® Technic

Spielzeugbausätze | Toy construction kits

LEGO GmbH
Technopark II

Werner-von-Siemens-Ring 14
85630 Grasbrunn
Germany
www.lego.com

Fußnote | Footnote
15.1: creative.lego.com/de-de/Products/Starter/5585.aspx
15.2: www.amazon.de/LEGO-8256-Technic-Go-Kart/dp/B001CQRS7E

Bild | Picture
50, 51: Bildrechte Fotografien | Photography rights Weller, Hinz, Krämer, Rajabi

16 Hybrid; OGEL
LEGO® Technic

Spielzeugbausätze | Toy construction kits

Hochschule Hannover
Simon Hellwig | Katharina Krämer | Birgit Weller
Fakultät III Information, Medien und Design
Expo Plaza 2
30539 Hannover
Germany
www.fakultaet3.fh-hannover.de
www.design-studenten.de

LEGO GmbH
Technopark II
Werner-von-Siemens-Ring 14
85630 Grasbrunn

Germany
www.lego.com

Bild | Picture
52, 53: Rendering Simon Hellwig
54: Bildrechte Fotografien | Photography rights Weller, Hinz, Krämer, Rajabi

17

Mutter; Depesche Vertrieb
Papa; Depesche Vertrieb

Geburtstagsgrußkarte | Birthday card

Depesche Vertrieb GmbH & Co. KG
Vierlander Straße 14
21502 Geesthacht
Germany
www.depesche.com

Lizensiert von b.sonders

Bild | Picture
55, 56: Bildrechte Fotografien | Photography rights Weller, Hinz, Krämer, Rajabi

18

Bosch private collection; Theo Klein
Bosch mini; Theo Klein

Spielzeug Bügeleisen | Toy iron
Spielzeug Kettensäge | Toy chainsaw

Theo Klein GmbH
Burgstraße 14
76857 Ramberg
Germany
www.my-klein-toys.de

Robert Bosch GmbH
Robert-Bosch-Platz 1
70839 Gerlingen-Schillerhöhe
Germany
www.bosch.com
www.bosch-pt.de/pt/de/boschmini

Fußnote | Footnote
18.1: www.kittoys.de/Hausgeraete/Buegeleisen-Bosch-Mini-Theo-Klein.html
18.2: Pressemitteilung www.bosch-presse.de/TBWebDB/de-DE/PressText.cfm?&Search=1&id=1066

Bild | Picture
57, 58, 60: Bildrechte Fotografien | Photography rights Weller, Hinz, Krämer, Rajabi
59: Werbedarstellung Robert Bosch GmbH

19 Pliers Combo (Pink); Tomboy Tools
Zangen; Wiha

Zangen | Pliers

Tomboy Tools, Inc.
2650 W. 6th Avenue
Bldg. E
Denver, CO 80204
USA
www.tomboytools.com

Wiha Werkzeuge GmbH
Obertalstraße 3-7
78136 Schonach
Germany
www.wiha.com

Fußnote | Footnote
19.1: www.tomboytools.ca/about.php

Bild | Picture
61, 62, 63, 64: Bildrechte Fotografien | Photography rights Weller, Hinz, Krämer, Rajabi

20 Vileda junior; Theo Klein
Bosch mini; Theo Klein

Spielzeug Besenwagen | Toy cleaning trolley
Spielzeug Werkbank | Toy work-station

Theo Klein GmbH
Burgstraße 14
76857 Ramberg
Germany
www.my-klein-toys.de

Vileda GmbH
Im Technologiepark 19
69469 Weinheim
Germany
www.vileda.de

Robert Bosch GmbH
Robert-Bosch-Platz 1
70839 Gerlingen Schillerhöhe
Germany
www.bosch.com
www.bosch-pt.de/pt/de/boschmini

Fußnote | Footnote
20.1: Philosophie der Theo Klein GmbH, www.my-klein-toys.de
20.2: Pressemitteilung, www.vileda.de/de/products/download/17166

Bild | Picture
65, 66, 67, 68: Bildrechte Fotografien | Photography rights Weller, Hinz, Krämer, Rajabi

21 Classic Alox; Victorinox
Handyman; Victorinox

Taschenmesser | Penknives

Victorinox AG
Schmiedgasse 57

6438 Ibach-Schwyz
Switzerland
www.victorinox.com

Fußnote | Footnote
21.1: www.funktionelles.de/Messer-Tools/Taschenmesser/Victorinox-Limited-Edition/Victorinox-Alox-Edition-orange.htm?shop=life&SessionId=&a=article&ProdNr=55648&t=31118&c=42551&p=42551
21.2: www.messerland.de/produkte/info/208/Handyman-1-3773.html/XTCsid/8a0d3ad07ec5300805898ce881bba3e2

Bild | Picture
69, 70, 71: Bildrechte Fotografien | Photography rights Weller, Hinz, Krämer, Rajabi

22 Air Vignon; menu
Leverpull LM-200; Screwpull

Korkenzieher | Corkscrews

MENU A/S
Kongevejen 2
3480 Fredensborg
Denmark
www.menu.as

Le Creuset GmbH
Einsteinstraße 44
73230 Kirchheim unter Teck
Germany
www.lecreuset.de/Produkte-/Screwpull

Fußnote | Footnote
22.1: www.menu-shop.de/menu-vignon-korkenzieher2.html
22.2: www.kochform.de/Screwpull-Korkenzieher-Leverpull-LM-250.htm

Bild | Picture
72, 73, 74, 75: Bildrechte Fotografien | Photography rights Weller, Hinz, Krämer, Rajabi

23 Storchschere; Wasa Solingen
Arbeitsschere; Forum

Stickschere | Embroidery scissors
Arbeitsschere | Working scissors

WASA Waldmin & Saam GmbH & Co. KG
Gottlieb-Heinrich-Str. 1-7
D-42719 Solingen
Germany
www.waldmin-saam.de

Union Werkzeug-Handel
E/D/E Einkaufsbüro Deutscher Eisenhändler GmbH
EDE Platz 1
42389 Wuppertal
Germany
www.union-werkzeuge.de/forum

Bild | Picture
76, 77: Bildrechte Fotografien | Photography rights Weller, Hinz, Krämer, Rajabi

24 Pretty Useful Tools™ Hammer; Picard

Werkzeugset | Tool-Kit
Latthammer, DIN 7239

Wild and Wolf Ltd

The Icehouse
124 Walcot Street
Bath BA1 5BG
United Kingdom
www.wildandwolf.com

Victoria and Albert Museum
V&A South Kensington
Cromwell Road
London SW7 2RL
United Kingdom
www.vam.ac.uk

Joh. Hermann Picard GmbH & Co.KG
Rottsiepen 15
D-42349 Wuppertal
Germany
www.picard-hammer.de

Fußnote | Footnote
24.1: www.edelight.de/i/werkzeugset-tool-kit-fuer-frauen-von-peter

Bild | Picture
78, 79, 80, 81: Bildrechte Fotografien | Photography rights Weller, Hinz, Krämer, Rajabi

25 sixx; fem Media
DMAX; Discovery Communications Deutschland

Fernsehsender | Television channels

fem Media GmbH
Ein Unternehmen der ProSiebenSat.1 Media AG
Tassiloplatz 7
81541 München
Germany
www.fem-media.com
www.sixx.de

Discovery Communications Deutschland GmbH & Co. KG
Maximilianstr. 13
80539 München
Germany
www.dmax.de

Fußnote | Footnote
25.1: www.sixx.de/online-tv
25.2: www.dmax.de

Bild | Picture
82: Werbedarstellung fem Media GmbH
83: Werbedarstellung Discovery Communications Deutschland GmbH & Co. KG

26 Kadett 912 Ideal Westernland; J.G.Schrödel
VIP Ideal Westernland; J.G.Schrödel

Spielzeugpistolen | Toy guns

Heinrich Bauer GmbH & Co.KG
Hans-Bunte-Strasse 2
90431 Nuernberg
Germany
www.sohni-wicke.de/de/produkte-alle-produkte-der-heinrich-bauer-group/Schrödel-Welt

Bild | Picture
84, 85, 86, 87: Bildrechte Fotografien | Photography rights Weller, Hinz, Krämer, Rajabi

27 Ergomat; Fiskars

Schere | Scissors

Fiskars Corporation
Hämeentie 135 A
P.O. Box 130
FI-00561 Helsinki
Finland
www3.fiskars.com

Fußnote | Footnote
27.1: www3.fiskars.com
27.2: www.manufactum.de/Produkt/193828/1438201/Schere-Ergomat.html

Bild | Picture
88, 89: Bildrechte Fotografien | Photography rights Weller, Hinz, Krämer, Rajabi

28 2911 Softouch®; Fiskars

Mehrzweckschere | Multi-purpose scissors

Fiskars Corporation
Hämeentie 135 A
P.O. Box 130
FI-00561 Helsinki
Finland
www3.fiskars.com

Fußnote | Footnote
28.1: www3.fiskars.com
28.2: www.fiskars-basteln.de/produkte.php?id=1

Bild | Picture
90, 91: Bildrechte Fotografien | Photography rights Weller, Hinz, Krämer, Rajabi

29 IXO; Bosch

Akku-Schrauber | Rechargeable screwdriver

Robert Bosch GmbH
Robert-Bosch-Platz 1
70839 Gerlingen-Schillerhöhe
Germany
www.bosch.com

Fußnote | Footnote
29.1: www.bosch-presse.de/presseforum/details.htm?txtID=3278&tk_id=112

Bild | Picture
92, 93: Bildrechte Fotografien | Photography rights Weller, Hinz, Krämer, Rajabi

30 Ergo; Bahco

Ergonomische Schraubendreher | Ergonomic screwdriver

SNA Europe Group
BP 20104 Eragn
95613 Cergy Pontoise
Paris
France

www.snaeurope.com
www.bahco.com

Bild | Picture
94, 95: Bildrechte Fotografien | Photography rights Weller, Hinz, Krämer, Rajabi

31 WMF 1; WMF

Kaffeepadautomat | Coffee pad machine

WMF Württembergische Metallwarenfabrik Aktiengesellschaft
Eberhardstraße
73309 Geislingen/Steige
Germany
www.wmf.de

Fußnote | Footnote
31.1: www.wmf.de/kaffee-tee/kaffeepadmaschinen_82511373.html?sid=F4D60838-80DA-4606-84C8-C1345DE7AB74
www.wmf.com/coffee-tea/coffee-pad-machine_82511373.html?sid=F0F759AE-F07D-4A9C-BD95-D751934ED683

Bild | Picture
96, 97, 98: Bildrechte Fotografien | Photography rights Weller, Hinz, Krämer, Rajabi

32 Cookware, Royal vkb

Kochgeschirr | Cooking pans

bv Koninklijke van Kempen & Begeer
Royal VKB
P.O.BOX 170

2700 AD ZOETERMEER
The Netherlands
www.royalvkb.com

Fußnote | Footnote
32.1: www.royalvkb.com/productdet.asp?catid=1212#

Bild | Picture
99, 100: Bildrechte Fotografien | Photography rights Weller, Hinz, Krämer, Rajabi

33 brand eins; brand eins Verlag

Wirtschaftsmagazin | Business magazine

brand eins Medien AG
brand eins Redaktions GmbH & Co. KG
Speersort 1
20095 Hamburg
Germany
www.brandeins.de

Fußnote | Footnote
33.1: www.brandeins.de/unternehmen/geschichte-philosophie.html

Bild | Picture
101, 102: Bildrechte Fotografien | Photography rights Weller, Hinz, Krämer, Rajabi

34 Rollbag; Authentics

Kulturtasche | Toiletries bag

Authentics GmbH
Am Ölbach 28
33334 Gütersloh
Germany
www.authentics.de

Fußnote | Footnote
34.1: www.authentics.de/shop/index.php/catalog/product/view/id/843/s/rollbag-l/category/5043

Bild | Picture
103: Bildrechte Fotografien | Photography rights Weller, Hinz, Krämer, Rajabi

35 Funnel; Normann Copenhagen

Haushaltstrichter | Household funnel

Normann Copenhagen
Østerbrogade 70
2100 Copenhagen
Denmark
normann-copenhagen.com

Fußnote | Footnote
35.1: normann-copenhagen.com

Bild | Picture
104, 105, 106: Bildrechte Fotografien | Photography rights Weller, Hinz, Krämer, Rajabi

Impressum
Imprint

Impressum | Imprint

Herausgeber | Publisher
Birgit Weller, Katharina Krämer

Das Projekt wurde gefördert durch den Gleichstellungsfond der Hochschule Hannover sowie die Hochschule für Technik und Wirtschaft Berlin. | The project is sponsored by the equalisation fund of the Hochschule Hannover University of Applied Sciences and Arts and the HTW University of Applied Science Berlin.

Verlag | Publisher
Blumhardt Verlag
Hochschule Hannover
Blumhardtstraße 2
30625 Hannover
Deutschland
http://www.blumhardt-verlag.de

Erscheinungsjahr 2012 | Year of publication 2012
ISBN 978-3-932011-85-6

Projektleitung | Project management
Birgit Weller

Konzeption und Redaktion | Concept and editor
Birgit Weller, Katharina Krämer

Wissenschaftliche Beratung | Scientific advice
Katrin Hinz

Autorenteam | Author team
Birgit Weller, Katharina Krämer, Katrin Hinz, Neda Rajabi

Lektorat | Editorial office
Marie Kuprat, Elisa Bader

Übersetzung | Translation
High Tech Translations, Hannover

Produktfotografien und Bildbearbeitung | Product photographs and image processing
Neda Rajabi, Andrea Barnett, Dennis Michaelis, Jonas Kakoschke, Katharina Krämer

Grafisches Konzept und Design | Graphic concept and design
Katharina Krämer, Neda Rajabi, Andrea Barnett, Dennis Michaelis
Layout und Satz | Layout and composition: Katharina Krämer
Designberatung | Design consulting: Sebastian Pilzner

Umschlaggestaltung | Cover design
Birgit Weller

Reinzeichnung und Druckvorbereitung | Final artwork and pre-press operations
Katharina Krämer

Druck und Bindung | Printing and binding
Ruksaldruck GmbH+Co.KG, Berlin

Papier | Paper
Umschlag 300g/m^2 Algro design
Inhalt 170g/m^2 Hello Silk

Schriften | Fonts
Scala, Scala Sans, The Sans

Redaktionsschluss | Editorial deadline
Dezember 2011

Die vorliegende Publikation ist das Ergebnis eines Forschungsprojektes der Hochschule Hannover in Kooperation mit der HTW Berlin, unterstützt aus Mitteln des Frauenförderfonds der Hochschule Hannover und dem Kompetenzfeld Universal Design Thinking der HTW Berlin. | This publication is the product of a joint research project of the Hochschule Hannover University of Applied Sciences and Arts and the HTW University of Applied Science Berlin, with funding from the Hochschule Hannover Frauenförderfund (Women's Promotion Fund) and the HTW Berlin KUD Universal Design Thinking competence field.

Die gleichnamige Ausstellung wurde erstmals im Rahmen der Konferenz *Die Diagonale – Universal Design Thinking* im November 2011 in Berlin der Öffentlichkeit präsentiert. | The exhibition *Die Diagonale – Universal Design Thinking* was first shown to the public in November 2011 in Berlin at the conference of the same name.

Prof. Birgit Weller
Hochschule Hannover | use-identity & design

Birgit Weller ist Produktdesignerin und seit 1994 Professorin für Industrial Design | Produktdesign in Hannover mit den Lehrgebieten: Entwurf, Entwurfsprozesse, Projektentwicklung und Präsentation. Sie initiierte und leitete vielfältige Forschungsvorhaben im In- und Ausland zu den Themen: Universal Design Thinking, Gender Codes im Design und Farbe im interkulturellen Kontext. Derzeit bereitet sie den Aufbau eines Universal Design Thinking Masters der Hochschule Hannover mit dem National Institut of Design sowie der HTW vor.

Nach dem Diplom 1985 an der Kunsthochschule Berlin arbeitete sie als Designerin bei LEW/AEG Transportation. Seitdem ist sie in den Bereichen Transportation-, Investitions- und Konsumgüterdesign, Packaging, Ausstellungsgestaltung und strategische Designberatung für weltweit agierende Unternehmen tätig. Sie arbeitet seit 2005 im internationalen Netzwerk INAREA und ist Mitbegründerin des Berliner Büros use: identity and design network. Sie agiert u.a. in Indien, Frankreich, Finnland, China, Niederlande. Seit 2002 ist sie im Vorstand des Internationalen Designzentrums Berlin und Mitinitiatorin den Kompetenzfeldes Universal Design Thinking – eine Kooperation der HTW mit der HSH. ■

Prof. Birgit Weller
Hochschule Hannover | use-identity & design

Birgit Weller is a product designer and professor of industrial design | product design in Hannover since 1994 covering the areas of: design, design process, project development and presentation. She has initiated and headed diverse research projects, both in Germany and abroad, on the subjects of: universal design thinking, gender codes in design and colour in an intercultural context. She is currently in the process of establishing a Universal Design Thinking masters at Hochschule Hannover together with the National Institute of Design and HTW Berlin.

After graduating (diploma) in 1985 at the Kunsthochschule Berlin (Berlin School of Art) she worked as a designer at LEW/AEG Transportation. Since that time she has taken on tasks in the fields of transportation, investment and consumer goods design, packaging, exhibition design and strategic design consultation for globally active companies. She joined the international INAREA network in 2005 and is co-founder of the Berlin-based design consultancy »use: identity and design network«. She has worked in many countries including India, France, Finland, China and the Netherlands. She has chaired the International Design Centre Berlin since 2002 and is co-initiator of the competence field Universal Design Thinking – in collaboration with HTW Berlin and Hochschule Hannover. ■

Katharina Krämer
Hochschule Hannover | die prototypen

Katharina Krämer ist Produktdesignerin und seit 2007 als freiberufliche Mitarbeiterin im Studiengang Produktdesign an der Hochschule Hannover tätig. Neben der Forschungsarbeit sowie Lehrtätigkeiten in den Forschungsprojekten *Genderspezifische Gestaltung von Produkten und Nutzungskonzepten und Universal Design Adaption* arbeitete sie im Lehrgebiet Entwurf. Im Jahr 2010 verbrachte sie einen Forschungsaufenthalt mit Lehrtätigkeit zum Thema Universal Design und Simple Product Design am National Institut of Design, India.

Nach dem Diplom 2007 an der Fachhochschule Hannover arbeitete sie als freiberufliche Designerin für verschiedene deutsche Unternehmen. Seit 2008 ist sie in den Bereichen Konzeptenwicklung, Konsumgüterdesign, Packaging Design und Corporate Design im eigenen Hannoveraner Büro die prototypen tätig, das sich aktuell zu einem Netzwerk umformiert. ■

Katharina Krämer
Hochschule Hannover | die prototypen

Katharina Krämer is a product designer and since 2007 freelance associate in the product design course at Hochschule Hannover. In addition to research work and teaching in the research projects *Gender-specific design of products and usage concepts* and *Universal design adaptation* she is also active in the field of design teaching. In 2010 she spent time researching the teaching of the subject of universal design and simple product design at the National Institute of Design, India.

Following her graduation (diploma) in 2007 at the Hochschule Hannover, she has worked as a freelance designer for various German companies. In 2008 she established her own Hannover-based design consultancy, die prototypen, where her activities include concept development, consumer goods design, packaging design and corporate design, whereby her office is currently being transformed into a network configuration. ■

Prof. Katrin Hinz
Hochschule fur Technik und Wirtschaft Berlin |
use-identity & design

Katrin Hinz ist Architektin und seit 1994 als Gründungsprofessorin im Studiengang Kommunikationsdesign in Berlin tätig, mit den Lehrgebieten: 3-D-Entwurf, Farbdesign, Orientierungssysteme und Verpackungsdesign. Sie initiierte und leitete vielfältige Forschungsvorhaben im In- und Ausland zu den Themen: Universal Design Thinking, eVideo, Design-Management und Farbe im interkulturellen Kontext. Seit 2011 baut sie als Gründungsprofessorin den neuen Bachelor-Studiengang Industrial Design an der HTW auf. Derzeit bereitet sie als Mitinitiatorin den Aufbau eines Universal Design Thinking Masters mit der der Hochschule Hannover und mit dem National Institut of Design vor.

Nach dem Diplom 1981 an der Kunsthochschule Berlin arbeitete sie als angestellte Architektin und als freie Bühnenbildnerin. Darauf aufbauend ist sie seit 1990 in den Bereichen Ausstellungsgestaltung, Event-Design, Packaging und strategische Designberatung für weltweit agierende Unternehmen tätig. Sie arbeitet seit 2005 im internationalen Netzwerk INAREA und ist Mitbegründerin des Berliner Büros use: identity and design network. Sie agiert u.a. in Ägypten, Indien, und den Niederlanden. Als Mitbegründerin des Designerinnen Forum war sie Mitglied des Gründungsvorstandes und langjähriges Mitglied des Internationalen Fachbeirats des Internationalen Designzentrums Berlin. Sie ist die Initiatorin ▸

Prof. Katrin Hinz
Hochschule fur Technik und Wirtschaft Berlin | use-identity & design

Katrin Hinz is an architect and one of the founding professors of the communications design course of studies in Berlin where she has lectured since 1994 covering the fields of: 3D design, colour design, orientation systems and packaging design. She has initiated and headed many research projects both at home and abroad on the subjects of: Universal Design Thinking, eVideo, design management and colours in an intercultural context. Since 2011, and as one of the founding professors, she is involved in establishing a new bachelor course of studies in industrial design at the HTW Berlin. She is currently a co-initiator in setting up the Universal Design Thinking masters course at Hochschule Hannover in cooperation with the National Institute of Design.

Following her graduation (diploma) in 1981 at the Kunsthochschule Berlin (Berlin School of Art), she worked as an architect and a stage designer. Since 1990 she has worked in the fields of exhibition design, event design, packaging and strategic design consultation for various globally active companies. A member of the international network INAREA since 2005, she is co-founder of the Berlin-based »use: identity and design network« design consultancy. She has worked in various countries including Egypt, India and the Netherlands. As a co-founder of the Designerinnen Forum she was a member of the founding executive committee and long-term ▸

und Sprecherin des Kompetenzfeldes Universal Design Thinking an der HTW. Von 2002 bis 2004 verantwortete sie als Vizepräsidentin für Forschung und Wissenstransfer maßgeblich die Standortkonzentration der HTW Berlin und ist seit 2008 Gastprofessorin und Vorsitzende des Beirates der Faculty of Applied Art and Science an der German University in Cairo. ■

member of the international advisory board to the International Design Centre in Berlin. She is the initiator and spokesperson of the competence field Universal Design Thinking at the HTW. From 2002 through 2004 she was vice-president of research and knowledge transfer and principally responsible for the site-concentration of HTW Berlin and since 2008 is guest professor and advisory board chairman to the Faculty of Applied Art and Science at the German University in Cairo. ■

Blumhardt Verlag
Publishing House

blumhardt verlag hannover

Blumhardtstraße 2, 30625 Hannover
Telefon: (0511) 9296-3162, Fax: (0511) 9296-3165
E-Mail: blumhardt-verlag@fh-hannover.de
www.blumhardt-verlag.de

... und Design ist doch eine Wissenschaft! Auf dem Wege zu einer akzeptierten Designwissenschaft, die bezugswissenschaftliche Fachbereiche nutzt, um auf wissenschaftlicher Ebene vergleichbar zu anderen Disziplinen im Diskurs anschließen zu können, ist dieses Buch ein weiterer Meilenstein: die strukturierte Promotion im Fach Design.

Marina-Elena Wachs,
Dorothee Weinlich
Promovieren im Design –
ein Kinderspiel?
Mit Ausblick für Fachhochschulen,
mit internationalen Vergleichsstudien
2011, 141 S., ISBN 978-3-932011-84-9,
Euro 20,00

Schriftenreihe der Fakultät V – Diakonie, Gesundheit und Soziales der Fachhochschule Hannover

Die Schriftenreihe der Fakultät V – Diakonie, Gesundheit und Soziales der Fachhochschule Hannover erschien bis Band 13 unter dem Titel Schriftenreihe der Evangelischen Fachhochschule Hannover.

5 Ulrike Mattke
Integration von Kindern mit Behinderung in der Landeshauptstadt Hannover.
Ein Forschungsbericht
2006, 175 S., ISBN 978-3-932011-65-8, Euro 8,00

6 M. Karl-Heinz Lehmann (Hrsg.)
Recht sozial.
Rechtsfragen der Sozialen Arbeit. 2., erweiterte Auflage
2002, VIII, 521 S., ISBN 978-3-932011-26-9, Euro 5,00

12 Gunter A. Pilz, Henning Böhmer (Hrsg.) (Hrsg.)
Wahrnehmen – Bewegen – Verändern.
Beiträge zur Theorie und Praxis sport-, körper- und bewegungsbezogener sozialer Arbeit
2002, 372 S., ISBN 978-3-932011-39-9, Euro 5,00

13 Thomas Grosse, Rosemarie Lüters
Grenzgänger konkret. Evaluation eines niedersächsischen Hip-Hop-Projekts
2007, 93 S., ISBN 978-3-932011-68-9, Euro 6,00

15 Dörte Detert
Gemeinsame Erziehungsverantwortung von Familien und Lehrkräften.
Studie zur Kooperationszufriedenheit am Beispiel der Primarstufe in Hannover und Liverpool
2007, 142 S., ISBN 978-3-932011-73-3, Euro 9,80

18 Elsbeth Krieg, Birgit Meinig, Simone Wustrack
Von der Kindertagesstätte zum Familienzentrum.
Entwicklungen und Herausforderungen für die Praxis
2010, 151 S., ISBN 978-3-932011-82-5, Euro 11,00

19 Rolf Balgo (Hrsg.)
Systemtheorie – eine hilfreiche Perspektive für Behinderung, Gesundheit und Soziales?
Dokumentation der Fachtagung »Systemtheorie – eine hilfreiche Perspektive für Behinderung, Gesundheit und Soziales?« vom 29.-30.10.2009 der Abteilung Heilpädagogik, Fakultät V – Diakonie, Gesundheit und Soziales der Fachhochschule Hannover
2010, 177 S., ISBN 978-3-932011-83-2, Euro 11,80

20 Rosemarie Lüters, Joachim Romppel, Maren Schreier
Fremdheit überwinden – Zusammenarbeit gestalten.
Ein Praxisleitfaden für die Arbeit mit Müttern und Vätern in Kindertageseinrichtungen
2011, 99 S., ISBN 978-3-932011-80-1, Euro 10,00

Quellen und Forschungen zum evangelischen sozialen Handeln
Herausgegeben von Martin Cordes und Rolf Hüper und Simone Wustrack

14 Martin Cordes, Heinrich Grosse (Hrsg.)
»Kirche, guck über deine Mauern!«.
Interviews zur Entwicklung des hannoverschen Kirchlichen Dienstes in der Arbeitswelt (KDA) nach 1945
2002, 323 S., ISBN 978-3-932011-41-2, Euro 5,00

15 Martin Cordes, Jürgen Filbir, Inge Petersen
Schule – Jugendarbeit – Kirche.
Konzepte und Projekte
2003, 204 S., ISBN 978-3-932011-42-9, Euro 5,00

19 Doris Riemann
Wo die Lebenden den Toten begegnen...
Mit einem Geleitw. von Barbara Duden
2007, 127 S., ISBN 978-3-932011-69-6, Euro 10,90

22 Simone Wustrack
Dem Aufwachsen Tiefe geben
Untersuchung zur religionspädagogischen Theorie und Praxis im evangelischen Kindergarten
2009, 237 S., ISBN 978-3-932011-78-8, Euro 22,00

23 Heinrich Grosse
»Niemand kann zwei Herren dienen«
Zur Geschichte der evangelischen Kirche im Nationalsozialismus und in der Nachkriegszeit.
2. durchgesehene Auflage 2010, 249 S., ISBN 978-3-932011-77-1, Euro 11,80

24 Ralf Hoburg
Zwischen Ortsgemeinde und Tourismus
Der Markt der Kirchenführungen als Herausforderung für die Kirche
2009, 143 S., ISBN 978-3-932011-75-7, Euro 10,90

25 Martin Cordes (Hg.)
Durch die Ritzen der Mauer
Kontinuitäten, Brüche, Neuanfänge in kirchlichen Partnerschaften nach 1949
2011, 231 S., ISBN 978-3-932011-79-5, Euro 11,00

Masterarbeiten der Fakultät V – Diakonie, Gesundheit und Soziales der Fachhochschule Hannover

Ausgezeichnet mit dem Preis des Blumhardt Verlags für herausragende Masterarbeiten.

1 Tobias Piontek
Stigmatisierung im Erleben von Jugendlichen mit Erfahrungen in der Psychiatrie
Eine empirische Untersuchung
2009, 142 S., ISBN 978-3-932011-81-8, Euro 10,90

Herausgeber | Publisher
Birgit Weller, Katharina Krämer

Verlag | Publisher
Blumhardt Verlag
Hochschule Hannover
Blumhardtstraße 2
30625 Hannover
http://www.blumhardt-verlag.de

Erscheinungsjahr 2012 | Year of publication 2012
ISBN 978-3-932011-85-6

Redaktionsschluss | Editorial deadline
Dezember 2011